T0222164

# Introduction to Intelligent Systems in Traffic and Transportation

# Synthesis Lectures on Artificial Intelligence and Machine Learning

Editors
**Ronald J. Brachman,** *Yahoo! Research*
**William W. Cohen,** *Carnegie Mellon University*
**Peter Stone,** *University of Texas at Austin*

Introduction to Intelligent Systems in Traffic and Transportation
Ana L.C. Bazzan and Franziska Klügl
2013

A Concise Introduction to Models and Methods for Automated Planning
Hector Geffner and Blai Bonet
2013

Essential Principles for Autonomous Robotics
Henry Hexmoor
2013

Case-Based Reasoning: A Concise Introduction
Beatriz López
2013

Answer Set Solving in Practice
Martin Gebser, Roland Kaminski, Benjamin Kaufmann, and Torsten Schaub
2012

Planning with Markov Decision Processes: An AI Perspective
Mausam and Andrey Kolobov
2012

Active Learning
Burr Settles
2012

Computational Aspects of Cooperative Game Theory
Georgios Chalkiadakis, Edith Elkind, and Michael Wooldridge
2011

Representations and Techniques for 3D Object Recognition and Scene Interpretation
Derek Hoiem and Silvio Savarese
2011

A Short Introduction to Preferences: Between Artificial Intelligence and Social Choice
Francesca Rossi, Kristen Brent Venable, and Toby Walsh
2011

Human Computation
Edith Law and Luis von Ahn
2011

Trading Agents
Michael P. Wellman
2011

Visual Object Recognition
Kristen Grauman and Bastian Leibe
2011

Learning with Support Vector Machines
Colin Campbell and Yiming Ying
2011

Algorithms for Reinforcement Learning
Csaba Szepesvári
2010

Data Integration: The Relational Logic Approach
Michael Genesereth
2010

Markov Logic: An Interface Layer for Artificial Intelligence
Pedro Domingos and Daniel Lowd
2009

Introduction to Semi-Supervised Learning
XiaojinZhu and Andrew B.Goldberg
2009

Action Programming Languages
Michael Thielscher
2008

Representation Discovery using Harmonic Analysis
Sridhar Mahadevan
2008

Essentials of Game Theory: A Concise Multidisciplinary Introduction
Kevin Leyton-Brown and Yoav Shoham
2008

A Concise Introduction to Multiagent Systems and Distributed Artificial Intelligence
Nikos Vlassis
2007

Intelligent Autonomous Robotics: A Robot Soccer Case Study
Peter Stone
2007

Introduction to Intelligent Systems in Traffic and Transportation

Ana L.C. Bazzan and Franziska Klügl

ISBN: 978-3-031-00437-7     paperback
ISBN: 978-3-031-01565-6     ebook

DOI 10.1007/978-3-031-01565-6

A Publication in the Springer series
*SYNTHESIS LECTURES ON ARTIFICIAL INTELLIGENCE AND MACHINE LEARNING*

Lecture #25
Series Editors: Ronald J. Brachman, *Yahoo Research*
         William W. Cohen, *Carnegie Mellon University*
         Peter Stone, *University of Texas at Austin*
Series ISSN
Synthesis Lectures on Artificial Intelligence and Machine Learning
Print 1939-4608    Electronic 1939-4616

# Introduction to Intelligent Systems in Traffic and Transportation

Ana L.C. Bazzan
Universidade Federal do Rio Grande do Sul (UFRGS)

Franziska Klügl
Örebro University

*SYNTHESIS LECTURES ON ARTIFICIAL INTELLIGENCE AND MACHINE LEARNING #25*

# ABSTRACT

Urban mobility is not only one of the pillars of modern economic systems, but also a key issue in the quest for equality of opportunity, once it can improve access to other services. Currently, however, there are a number of negative issues related to traffic, especially in mega-cities, such as economical issues (cost of opportunity caused by delays), environmental (externalities related to emissions of pollutants), and social (traffic accidents). Solutions to these issues are more and more closely tied to information and communication technology. Indeed, a search in the technical literature (using the keyword "urban traffic" to filter out articles on data network traffic) retrieved the following number of articles (as of December 3, 2013): 9,443 (ACM Digital Library), 26,054 (Scopus), and 1,730,000 (Google Scholar). Moreover, articles listed in the ACM query relate to conferences as diverse as MobiCom, CHI, PADS, and AAMAS. This means that there is a big and diverse community of computer scientists and computer engineers who tackle research that is connected to the development of intelligent traffic and transportation systems. It is also possible to see that this community is growing, and that research projects are getting more and more inter-disciplinary. To foster the cooperation among the involved communities, this book aims at giving a broad introduction into the basic but relevant concepts related to transportation systems, targeting researchers and practitioners from computer science and information technology. In addition, the second part of the book gives a panorama of some of the most exciting and newest technologies, originating in computer science and computer engineering, that are now being employed in projects related to car-to-car communication, interconnected vehicles, car navigation, platooning, crowd sensing and sensor networks, among others. This material will also be of interest to engineers and researchers from the traffic and transportation community.

# KEYWORDS

intelligent transportation systems, traffic modeling, traffic simulation, advanced traveler information systems, traffic control, traffic assignment, traffic management, route choice, routing, route guidance, driver assistance systems, car to car communication, artificial intelligence, machine learning, reinforcement learning, swarm intelligence, multiagent systems

# Contents

# Preface

Over the last decades, there has been an increasing interest in aspects that lie in the intersection of computer science and traffic engineering. This is not surprising: in the introductory chapter of this book we provide several figures about the importance of transportation to the economy and daily life of the modern citizen, especially in urban environments. Just to mention one of those figures, congestion is now mentioned as one of the major problems in various parts of the world. This naturally brings in the issue of optimization, a subject that is dear to computer scientists. Besides, the diffusion of the Internet of Things and Mobility Internet promises to reduce the gap between computer engineering and traffic engineering. Indeed, it is predicted that we will see a surge in shipments of car wi-fi systems. This opens up the possibility for completely new ways to manage mobility of people and goods, as well as to optimize traffic flows, just to mention two obvious examples of new possibilities. In fact, the so-called smart mobility is one of the pillars of research projects related to smart or intelligent cities. Smart mobility aims at connecting the use of technology (Internet, sensors, mobile devices) to gather and integrate information in order to improve the efficiency of the transportation system (e.g., reducing congestion, optimize infrastructure, reduce costs, accidents, emissions, and increase the comfort of the user).

Given this panorama, the goal of the present book is to give a broad introduction into the relevant concepts related to traffic and transportation systems, targeting researchers and practitioners from computer science and information technology.

## READERSHIP AND ROADMAP

This book is directed towards computer science students, practitioners, and researchers who want to learn the basics on traffic systems and how intelligent methods from AI and other computer science areas can be used for solving traffic-related problems. Understanding the book content requires basic computer science knowledge. On the one hand, the book is mainly interesting for people without deeper knowledge on traffic systems. For these, Chapters 2, 3, and 4 give the introduction to concepts related to traffic infrastructure, demand, and demand assignment, respectively. Chapter 5 deals with the, often relegated, issue of acquisition of traffic data.

On the other hand, this book might also be interesting for people with a background in traffic engineering, who are interested in getting a perspective from the computer science and information technology (IT) points of view, especially when it comes to ITS (Intelligent Transportation Systems), which we introduce in Chapter 1. Besides, these readers may also find, in the last chapter, a discussion of recent trends and use of new technologies that appear with the possibility of electro-mobility and car-to-car communication.

Chapters 6, 7, and 8 are directed to both computer scientists and traffic engineering researchers and practitioners; they discuss issues that include microscopic and agent-based modeling and simulation, intelligent control of traffic signals, and in-vehicle route guidance, among others.

## LITERATURE COVERING RELATED AREAS

The present textbook focuses on vehicular traffic. Despite this topic being closely related to others, such as simulation of pedestrian movements, logistics, public transport, air traffic control, and others, we intentionally do not cover these. The interested reader is referred to textbooks such as Morlok [1978] and Sussman [2000], or survey papers that focus on particular issues, such as agent-based management of logistics [Davidsson et al., 2005] or multiagent approach to manage air traffic control Agogino and Tumer [2012].

## PERSONAL PRONOUNS

We found that the sole use of "she" (an obvious choice given the authorship!) would be unfair. Therefore we use both "he" and "she" in a more or less random way.

Ana L.C. Bazzan and Franziska Klügl
December 2013

# Acknowledgments

We are grateful to Peter Stone and Mike Morgan for suggesting this volume.

We also would like to thank Andrew Koster for reading a draft version of the text. We are grateful to him and the anonymous reviewers, for carefully reading preliminary versions of this book and giving us helpful feedback regarding distribution of contents, style, and language, as well as for pointing to additional material.

Luiz H.D. Souza implemented the simulation of the circular road in SUMO, which was used to illustrate the concepts of shock wave and fundamental diagram. Gabriel de O. Ramos and Alessandro D. Vecchia have provided manual counting of vehicles.

Icons from OpenClipArt (http://openclipart.org/) were used in Figures 1.1 and 2.4.

Ana Bazzan gratefully acknowledges the partial funding of the Brazilian agencies CNPq and FAPERGS, as well as of the German Foundation Alexander von Humboldt.

Franziska Klügl gratefully acknowledges the partial funding by VINNOVA in the VINNMER schema, as well as by the Swedish Foundation for International Cooperation in Research and Higher Education (STINT) for supporting the cooperation with Ana Bazzan.

Ana L.C. Bazzan and Franziska Klügl
December 2013

# List of Symbols

| | |
|---|---|
| $a$ | acceleration |
| $C$ | capacity |
| $\Delta_t$ | time interval |
| $k$ | density |
| $K$ | countings (of number of vehicles) |
| $N_x$ | number of vehicles (vehicle count) at $t_i$ in $\Delta_x$ |
| $N_t$ | number of vehicles (vehicle count) at $x_i$ in time interval $\Delta_t$ |
| $q$ | flow |
| $s$ | distance |
| $v$ | speed |
| $V$ | volume |
| $\bar{v}$ | average speed |
| $\theta$ | offset (progressive systems) |
| $t$ | time |
| $T$ | number of trips |

# CHAPTER 1

# Introduction

## 1.1  THE IMPORTANCE OF TRANSPORTATION

Recently, the number of city dwellers has surpassed 50% of the total population of the planet. According to the United Nations, for 2050, it is forecasted that around two-thirds of us will be living in urban agglomerations. As a consequence, the number of large metropolitan areas with more than ten million inhabitants is increasing rapidly, with the number of the so-called mega-cities now (2013) at more than 20. For comparison, in 1950 there were 83 cities with populations exceeding *one million*, with New York being the sole city with population above ten million.

The increase in the number of mega-cities has huge consequences to traffic and transportation. In fact, the second half of the last century has seen the beginning of the phenomenon of traffic congestion. This has occurred because the demand for mobility in our society has increased constantly. Besides, transportation and economic growth are closely tied: according to studies conducted by the European Union (Delphy Study on Future and Mobility), the volume of goods and people transported follows the growth of the GDP. The major problem with this is that the increase in transportation volume generates traffic congestion. The increase in capacity, if any, handles only a tiny part of such increase. The consequences are well known: in the last decade, mobility patterns have changed drastically and now congestion is mentioned as one of the major problems in various parts of the world, leading to a significant decrease in the quality of life, especially in mega-cities of countries experiencing booming economies. This causes delays, air pollution, decrease in speed, and insatisfaction, which may lead drivers to risky maneuvers that reduce safety for pedestrians, as well as for other drivers.

Although the number of fatalities per distance traveled has been decreasing in the U.S. and Europe in the last decades (mostly due to improved security techniques developed by the car industry), fatality figures for the U.S. collected by the Department of Transportation (`http://www-fars.nhtsa.dot.gov`) show that in 2011 there were around 32,000 traffic-related deaths in the U.S. alone. Statistics from the European Union show that around 50,000 fatalities occur per year in Europe. Figures for BRICS and developing countries in general are not different. For example, approximately 30.000 fatalities and 320,000 injuries occur annualy in traffic accidents in Brazil.

Apart from human and material losses, there is the issue of impacts on the environment and on mobility patterns. In the U.S., it is estimated that travel delay reaches 4.8 billion hours, which puts the cost of urban congestion at 114 billion dollars. In 2007, delays caused by congestion have costed US$87.2 (50% more than in 1997). Americans travel 4.2 billion hours more than if there were no traffic jams, and consume 3–4 billion more gallons of fuel, with great negative impact on the environment. Still for the U.S., 20% of Gross National Product (GNP) is spent on transportation, of which about 85% on highway transportation (passenger and freight). There are 150 million automobiles (that means car ownership for 56% of the population) and these vehicles are driven an average of 10,000 miles per year for passenger cars and 50,000 miles per year for trucks on a highway system that comprises more than 4 million miles [Gartner et al., 2001].

Thus, it seems clear that there is an increase in transportation demand, even if this is not a general case.[1] Until some decades ago, this increase could be met by providing additional capacity. Nowadays this no longer seems to be economically or socially attainable or feasible. Other measures have been adopted in recent years, such as congestion charging in urban areas (London), restriction of traffic in the historical center (Rome, Paris, Amsterdam), rotating vehicles allowed to circulate in a given day (São Paulo, Mexico City), and many others that seek to impose a sustainable mobility. However, these measures are not popular among tax-payers. Thus, traffic engineering seeks to improve the existing infrastructure, without increasing the overall nominal capacity by means of an optimal utilization of the available capacity. This transcends pure technical issues: according to the keynote speaker of the IEEE 2011 forum on integrated sustainable transportation systems, Martin Wachs, "mobility is perhaps the single greatest global force in the quest for equality of opportunity" because it plays a role in offering improved access to other services.

## 1.2    TRANSPORTATION SYSTEMS

There is no simple definition of what constitutes a transportation[2] system. Various definitions can be found, depending on the context, perspective, and purpose of the specific technical literature. For instance, when talking about modes of transportation, each of these modes can be seen as a system (e.g., the transit system is commonly defined as a transportation system). Then, a transportation system can be seen as composed by a supply part (generally meaning the physical infrastructure that provides the capacity) and by a demand part. Finally, the scale of the system (e.g., urban, interurban, international) can also be used to characterize a transportation system.

The function of a transportation system is to provide for the movement of things (both inanimate objects as well as persons, animals, and plants) [Morlok, 1978]. This book focuses on transportation of persons, in specific situations, and the generation of traffic. Thus, we use the terms traffic system and transportation system interchangeably, even if this is not conceptually and generally correct.

The various components of a transportation system are road users (drivers, transit passengers, pedestrians, cyclists, motorcyclists); vehicles (both private and commercial), buses, and other

---

[1] According to the Texas Transportation Institute, in 2007, congestion in U.S. urban areas has declined. One possible explanation is the increase in fuel cost. This may hint that road pricing is a way to mitigate congestion, as it also increases travel costs.

[2] In this text we use the American English denomination: transportation; while Americans tend to use transport as a verb only, the British usually refers to public transport where Americans would use public transportation.

transit means; streets and highways; traffic control and conventions, laws, enforcement policies, lightning, etc.). While road users clearly relates to the demand, vehicles and transit in general can be seen as serving this demand. Streets, highways, and control devices relate to supply.

An important point when investigating transportation systems is that, frequently, they are mentioned as examples of complex systems. Indeed, a transportation system has at least three characteristics of complex systems: (i) it is composed of units (subsystems) that are coupled in ways that are not always perfectly known; (ii) its emergent behavior is difficult to predict, even when its components are themselves predictable; and (iii) small changes in inputs or parameters may produce large changes in behavior.

This last characteristic reminds us that transportation systems are dynamic systems. For instance, if we consider vehicular traffic as a many-particle system, its dynamics is often associated with the so-called spatio-temporal behavior. Its complexity is due to nonlinear interactions between: (i) travel decision behavior, which determines traffic demand; (ii) routing of vehicles in a traffic network; and (iii) traffic congestion occurrence within the network.

As seen in Figure 1.1, travel decision behavior determines travel demand. Routes are associated with the infrastructure (supply). However, traffic congestion may arise in the network restricting free flow travel, which then influences both travel decision behavior in general and selection of routes in particular. For example, if facing traffic jams, a user may decide to stay at home, postpone the trip, or select another transportation mode. This, of course, requires perception or awareness of the options. This represents the feedback loop from the user's perspective (left side of Figure 1.1). The other feedback loop (right side of the figure) arises from the system's perspective, as the system managers, traffic controllers, or urban planners acquire data about changes in demand, and, if possible, react with changes in the infrastructure or supply of transportation options. One consequence of these feedback loops is that demand causes traffic in the infrastructure, and more infrastructure is likely to attract more traffic, and so on. Therefore, both feedback loops permeate all transportation research. However, it is simpler to tackle each part separately. The system and the user loops can roughly be related to supply and demand respectively. This is the perspective we take in Chapters 2 and 3.

In the remainder of this chapter we briefly introduce notions of demand, which are later detailed in Chapter 3. Following this, we give discuss general practices for regulating the demand in order to achieve an efficient use of the infrastructure, anticipating the material that appears in Chapter 2. Section 1.3 then tries to convey a more recent view that aims at unifying these two perspectives under the single umbrella of intelligent transportation systems (ITS).

In traffic engineering, a common unit used to measure demand is number of vehicles or number of trips (by private vehicles or by other means of transportation), depending on the primary focus of the technical literature on transportation and traffic. Roess et al. [2004] points to the issue that the demand actors are far from homogeneous. This refers both to a heterogeneity in terms of what each actor demands from the system, as well as in terms of which capabilities and resources these actors have. Indeed, one can find elderly, as well as novice drivers, aggressive as

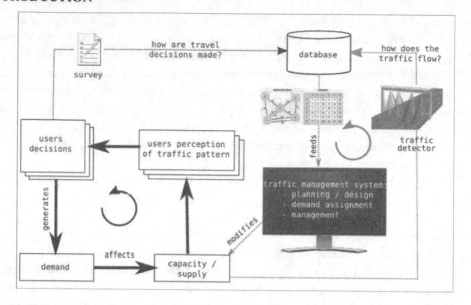

**Figure 1.1:** Transportation system with its main components and their relationship: user and system feedback loops.

well as timid drivers; drivers have different reaction times, vision characteristics; pedestrians walk with different speeds. All these issues must be taken into account when designing traffic systems. We return to this issue in Chapter 6, when we discuss the advantages of microscopic simulation and, in particular, of agent-based simulation.

Apart from posing challenges in modeling and simulation, the diversity/heterogeneity in road users' behaviors has a consequence that it is not easy to predict traffic streams. This observation is important because some models of traffic streams are based on fluid dynamics, among others. However, a flow of water through pipes can be exactly predicted, whereas this is not the case when dealing with road users.

There are indications that Romans and even more ancient civilizations had already established some forms of regulation, if not control, of traffic, thus addressing issues related to the infrastructure (supply). In modern times, traffic control has been traditionally achieved by employing traffic lights, as discussed in Chapters 2 and 7. However, we note that congestion can also be addressed by means of traffic regulations and policies. The former involves speed limits, on-ramp metering, bans on passing for slow vehicles and reversible lanes, among others. The latter, policies, can be implemented by modifying or even removing infrastructure elements, as a mean to create incentives to use different means of transportation, and dispersing rush hours. This distinction between regulations and policies is important because it has impact on what kind of tools have to be developed. While regulations can be analyzed with tools that simulate traffic flow (possibly at microscopic level), analysis of effects of new policies need tools that address the transportation network at planning or macroscopic level. This is discussed in Chapter 6.

## 1.3     INTELLIGENT TRANSPORTATION SYSTEMS

Apart from developing simulation tools, the role of computer scientists (and software developers in general) in traffic and transportation system is likely to grow in importance in the near future. Indeed, recent advances in information and communication technology open the opportunity not only to new forms of simulation and control, but also a new array of possibilities for improving the throughput, safety, and equity of using the infrastructure. Such new technologies are the basis of what is known as intelligent transportation systems (ITS). Since a great deal of ITS is about information, here lies a great opportunity for AI practitioners and computer scientists to work in multidisciplinary teams. This may be key in making more efficient use of the existing transportation system, as it is unlikely that it is going to expand significantly due to economic, environmental, and practical constraints.

There is no standard definition for ITS. Rather, this term is flexible and can be interpreted in a broad or narrow way. We remark that in Europe the term transport telematics is often used as synonym for ITS.

The main aim of ITS is to improve decision making, often in real time, by transport and traffic authorities, as well as by other users.

Depending on the country and context, ITS may refer to all modes of transportation or, in its most common definition, to road transportation. In this case, ITS can be seen as systems in which information and communication technologies (I&CT) are applied in areas related to the transportation network (for instance, infrastructure, vehicles and road users, traffic and mobility management, and the dynamic interaction among all these elements). Hence, ITS can be seen as a general term for the integrated application of communications, control, and information processing technologies to the transportation system. This way, ITS involves a broad array of techniques and approaches related to I&CT. Information is at the core of ITS. Thus, many ITS tools are dedicated to the collection, processing, integration and provision of information. These information is important not only for operators and traffic and transportation authorities, but also public and commercial transport providers, and individual road users. As mentioned, the aim is to allow more intelligent decision making, with the ultimate goals of making the transport system safer and more efficient, and to lead to a smarter use of the network.

Since the 1990s, several elements of what is now known as ITS have been defined, with emphasis on traffic surveillance, control, signal optimization, variable message signing (VMS), and simulation of traffic and transportation systems. The appearance of ITS is mostly credited to the advances in computing and communication. As computing costs decreased in the 1990s, it was possible to deploy microprocessors and more intelligence in traffic systems. More recently, advances in communication (satellites, cell phone, and other mobile devices) enables the use of navigation devices, now provided by several companies, together with commercial vehicle routing systems.

Therefore, typical computing-related issues and techniques are becoming common place in traffic and transportation systems, such as ubiquitous computing, internet of things, and cloud computing. This situation allows individuals (drivers, passengers, etc.) to have continuous and ubiquitous access to information, which is delivered both by the private sector, and also by the road users themselves (as seen by collaborative platforms such as Waze). This constitutes a clear change in paradigm if one thinks that a decade ago the main provider of traffic and transportation information was the public sector (traffic authority). This change in paradigm, in turn, means that these authorities have less and less control over the network since more informed individuals may continuously adapt their choices to fit their mobility goals. Open questions are what the consequences of such behavior are, and whether the main role of the public sector is going to be the enforcement of rules (now also facilitated by I&CT as automated surveillance systems such as cameras and radars that communicate with operation centers in real time). This is a controversial question. Some authors also ascribe a major role to the public sector, namely one related to control of demand by means of electronic road pricing (congestion toll) to influence temporal, spatial, and modal choices, as a way to make a better use of the network.

ITS generally encompasses five areas or systems:[3] (i) advanced transportation/traffic management systems (ATMS);[4] (ii) advanced traveler information systems (ATIS); (iii) advanced vehicle control and safety systems (AVCSS), sometimes also named advanced driver assistance systems (ADAS); (iv) advanced public transportation systems (APTS); and (v) commercial vehicle operation (CVO). In this book, although we focus on the former two—for they encompass most of the functions and services that are related to road transportation—we give a brief description of all of these systems.

ATMS aim at managing technologies related to traffic control devices, management of emergency situations, monitoring of emissions, and communication among the various parts of the system, such as traffic monitoring devices, traffic signal controllers, and other devices related to safety.

The goal of ATIS is to provide information to road users, transit users, and other participants of the transportation system in freeways, urban environments and all other roads. Such information is, in many cases, collected and processed by the ATMS and then broadcast using various media: radio, navigation devices, variable-message sign, etc., both for pre-trip as well as for en-route planning. Whereas ATMS refer primarily to infrastructure (supply), ATIS are directed at the users of the system (demand).

---

[3]It is important to make the point that there is no standard terminology in the area of ITS; these acronyms may appear differently in other chapters (as we tend to use the most common terminology in each context); we try to make the connections when they are necessary.

[4]In the literature, the "T" in ATMS refers to both transportation and traffic management, and less commonly, also to travel.

AVCSS's aim is to apply advanced technologies in vehicles and roads in order to reduce accidents and improve traffic safety. These systems include anti-collision warning and control, driving assistance, automatic lateral/longitudinal control, etc.

APTS apply the technology of ATMS, ATIS, and AVCSS in public transportation in order to improve the quality of service, and increase efficiency by means of automatic vehicle monitoring and e-ticketing.

Finally, CVO works like APTS but applied to commercial vehicle operations (automatic vehicle monitoring, fleet management, scheduling, and electronic payment).

In terms of applications of ITS, most of them focus on the kernel view of ITS: providing information to the user. Typical questions that arise are how drivers are rerouting in response to the information received, and how this affects the network load. A related question is how to estimate the actions of drivers based on their current location, the origin-destination matrix that underlies their preferred routes, and their decision rules, which are basically unknown.

With the dissemination of the use of smart-phones and other mobile devices, this is more and more a target of private firms, which now compete to get such information not only for traffic related applications but also for commercial purposes. Issues related to privacy and rights of collecting and disseminating data are currently open.

These issues relate to the dynamic assignment and routing problems that are discussed in Chapter 4: a large share of drivers tend to select routes based on their historical knowledge. If some events cause a disruption in the capacity of part(s) of the network, how do drivers reroute? A dynamic traffic assignment model has to anticipate what these drivers will do and, in some cases, advise them. This task of modeling and anticipation is also part of the ITS effort since it involves simulation that requires efficient computer models and hardware.

Other applications of ITS relate to traffic signals optimization using techniques from AI (e.g., fuzzy logic); ramp metering; emergency services; disaster management; enforcement of traffic laws; parking management; fleet management (including taxis); communication with drivers; electronic toll collection; electronic road pricing; mass transit smart cards; collecting weather data; and logistics.

Many of these applications generate an enormous load of data that can be used not only to improve the applications and optimize the operation of the businesses responsible for the production of such data (e.g., travel times of service vehicles), but also to apply data mining techniques to find out patterns about OD estimates, trends in land use and transit use, as well as to optimize traffic signals.

# CHAPTER 2

# Elements of Supply

In this chapter we describe the basic concepts related to the infrastructure of traffic and transportation system, which can be seen as the supply side. We start with definitions related to the network structure, then address elements of traffic streams and traffic flow theory. Following, we discuss traffic control measures.

## 2.1  TRAFFIC NETWORK STRUCTURE

Networks are useful to describe transportation systems from the point of view of physical components. They are composed by nodes (vertices) and links (edges). Nodes represent stations, junctions, etc. There is normally more than one way to travel between two given nodes, i.e., the network is redundant. Links represent roads, highways, rail lines, etc. Each link conducts flow (e.g., vehicles in one or two directions) and has a given nominal capacity. There are several ways to represent the cost of traversing a link, but usually the cost is a function of the capacity and its current occupancy. Costs can be explicit (tolls), implicit (travel time), or more abstract representations such as delay-cost functions as discussed in Section 4.4.

Each traffic authority has particular names for the various kinds of links but a widely accepted hierarchy is based on their capacity and function. Regarding capacity, according to Morlok [1978], one has residential streets emptying into collector streets, which empty into arterials that then empty into freeways. A functional definition of roadway is in terms of the tradeoff between access and speed. For instance, residential streets provide high access but slow speeds, freeways have very limited access but high speed, etc.

The task of describing traffic streams involves the measurement and analysis of parameters that broadly fall into two categories: macroscopic (volume, speed, and density of traffic streams) and microscopic (speed of individual vehicles, headway and gap or spacing). Before addressing details about these parameters, it is important to mention that the type of roadway plays a key role on the overall behavior of such parameters.

Roadways with uninterrupted traffic flow present no interruption such as intersections with traffic signals, stop or yield signs, etc. This is primarily the case with freeways. Here, the characteristics of the traffic streams are based mainly on the interaction among vehicles. It must be noted that the environment also plays a role (e.g., weather conditions). In roadways with interrupted flow, external interruptions such as traffic signals determine alternation of stop and go for road users, which creates platoons of them. Clearly, interrupted flows are more complex to model and represent since movements are periodically constrained (e.g., by a red signal).

Finally, it is worth mentioning that the aforementioned microscopic and, especially, macroscopic parameters are also closely related to performance measurements. In its most known variant—the one introduced in 1965 by the Highway Capacity Manual[1] (HCM)—the concept of level-of-service (LoS) aims at describing the quality of operations on a facility (freeway, arterial, etc.), under defined conditions. The LoS uses a scale from A (best) to F (worst). The HCM in its 2000 version [Tra, 2001] not only defines LoS but also gives precise measures of effectiveness. That edition of the HCM also defines performance measures such as average speed and density (for uninterrupted flow), and delay and average queue size (for signalized intersections).

## 2.2   TRAFFIC STREAMS: MACROSCOPIC AND MICROSCOPIC PARAMETERS

As mentioned, describing traffic streams involves a series of parameters that can be macroscopic or microscopic. Speed ($v$) is of course a microscopic property of each vehicle. Hence, an average speed value can be associated with the whole stream. Alternatively, a distribution of speed values can be used. Average or mean speed is often measured over all vehicles occupying a given *section* of the road in a given *time unit*. This is known as space-mean. Average or mean speed can also be measured by taking into account all vehicles passing a given *point* on a road or lane over some specified time period [Roess et al., 2004]. The latter is known as time-mean.

The most characteristic macroscopic parameters however are the traffic volume (traffic flow, flow rate) $q$, and the density $k$. The former (measured in number of vehicles or vehicles per unit time) is defined as the number of vehicles passing a point on a road, or a given lane or direction of a road, during a specified time interval. A related concept is the one of volume, which is normally used for daily or hourly volumes for peak hours, as well as averaged annual, daily, hourly volumes [Roess et al., 2004].

Density—macroscopic measure of the proximity of vehicles that are part of a stream—is defined as the number of vehicles occupying a given length of a road or lane, expressed as vehicles per length unit. The relationship of density with speed comes from the fact that drivers select speeds that are consistent with how close vehicles are [Roess et al., 2004]. Moreover, density is an important measure of the quality of traffic flow. Whereas density is difficult to measure, modern detectors can measure occupancy, which is the proportion of time that the detector is covered by a vehicle in a defined time period. Otherwise, density is computed as a function of flow and speed (see next section).

Besides speed, flow and density are also related to parameters that refer to individual vehicles. For instance, average density is related to spacing, i.e., the distance between two successive vehicles. Headway is defined as the time interval between successive vehicles, measured at a reference point, and is directly related to average flow.

---

[1]A publication of the U.S. Transportation Research Board (TRB, www.trb.org).

As mentioned, macroscopic parameters can be obtained from microscopic ones (such as vehicle positions and speeds) by local aggregation. The reverse operation, i.e., obtaining information about individual vehicles from macroscopic quantities by disaggregation, is more difficult.

## 2.3    TRAFFIC FLOW THEORY

After the classical work of Greenshields [1935], traffic flow models follow the research line around the so-called fundamental diagram of traffic flow (henceforth fundamental diagram for short). This diagram shows the relationship between the traffic stream parameters introduced in the previous section: flow (volume), density, and speed. For instance, the fundamental diagram reflects the obvious idea that the higher the density the lower the speed.

Typically, the fundamental diagram plots flow against density. However, because speed is determined by the slope of the secant at any point of corresponding to a given pair flow-density, the fundamental diagram illustrates the relationship among these three variables. Mostly, these relationships derive from empirical observation of real traffic, which then generate spatio-temporal patterns. Traffic flow models may consider either macroscopic or microscopic variables as input.

In Chapter 5, some methods for measuring these variables are discussed. Assuming that the number $N$ of vehicles passing a given location and their individual speeds can be measured, if a given time interval $\Delta_t$ (for instance 1 h) is considered, then the flow rate is given by Eq. 2.1. Then, the density $k$ (as mentioned, the number of vehicles per length unit as, e.g., 1 km) can be computed as in Eq. 2.2, where $\bar{v}$ is the average speed computed by dividing the summation of speeds of all vehicles that passed the given location during $\Delta_t$.

$$q = \frac{N}{\Delta_t} , \tag{2.1}$$

$$k = \frac{q}{\bar{v}} . \tag{2.2}$$

Dimensions of these fundamental quantities are as follows:

$$q \left[ \frac{\text{vehicles}}{\text{time}} \right] = k \left[ \frac{\text{vehicles}}{\text{distance}} \right] \times \bar{v} \left[ \frac{\text{distance}}{\text{time}} \right] .$$

The speed-density relationship reflects the behavior of drivers because these select their speeds primarily based on the perception of the proximity to other vehicles and on their concept about safe operation in given density conditions. Density is thus created by each driver's selection of speed (which, in congested regimes, is itself be a function of density). Both density and speed then define the rate of flow.

The behavior of drivers depend, among other factors, on how many vehicles ahead a driver perceives, and on how close they are. In the free flow regime, speed depends neither on traffic volume nor on density. In this regime, as $k \to 0$, $\lim_{k \to 0} \frac{d\bar{v}}{dt} \to 0$. On the other hand, when $\bar{v} = 0$,

$k = k_{max}$. For road traffic, $k_{max} = 150$ veh/km/lane is a rough guideline [Leutzbach, 1988]. We return to this subject in Section 6.1.2, where we introduce the car-following model.

From the flow-density diagram, the following quantities can be derived: (1) the free-flow speed (the asymptotic gradient of the curve when $k = 0$); (2) the actual average speed for a given $k$ (the slope $q(k)/k$ of the secant through $(0, 0)$ and $(k, q(k))$); and (3) capacity of the road or lane (the maximum value of $q(k)$).

Given that the three critical variables—speed, flow, and density—are related by Eq. 2.2, calibrating a relationship between any two of them determines the third. Thus, the number of combinations of values for the three variables is restricted. Figure 2.1 depicts the theoretical form of the relationship between these variables. Later we show an empirical curve for speed-flow.

At least theoretically, the relationship between density, speed, and flow rate is clear: as density increases, the average speed decreases, causing the flow rate to decrease as well. When speed is zero, then flow is zero as well, which can be seen by the theoretical fundamental diagram in Figure 2.1. From the plot of flow vs. density, we see that in the free flow regime, the slope of the tangent to the fundamental diagram at the origin determines $\bar{v}_{opt}$.

The fundamental diagram is used to characterize different flow regimes. A free flow regime is a regime in which each vehicle travels at the driver's desired speed (obviously observing road and vehicles constraints). According to Leutzbach [1988], this kind of flow only occurs if few vehicles are on the road and there are sufficient lanes to allow overtaking without delay. With increase in density of vehicles, these start to be unable to maintain their desired speed because there is a reduction in speed in order to fit the speed of slower vehicles ahead. This results in a continuous decrease in average speed. This regime is called partly constrained because not all vehicles are free to move at their desired speeds and/or to overtake. When unconstrained overtaking is no longer possible, vehicles start traveling in platoons whose speed is determined by the vehicle ahead. Also, several slow drivers will cause the platoons to break into even more platoons, defining a constrained regime of traffic.

The transition from partially constrained to constrained traffic is clearly noticed in the relationship between speed and flow (see, e.g., Figure 2.1). It occurs (at least theoretically) at the maximum of the curve for the fundamental diagram depicting flow rate $q$ vs. speed $\bar{v}$, where the variation of flow with speed approaches zero. The speed at this point is denoted $\bar{v}_{opt}$ or $\bar{v}_{crit}$. On the other hand, the free flow regime is the one in which changes in speed can be neglected, i.e., they do not constraint the movement at desired speed of other vehicles. Some authors just consider two regimes: free flow and congested. For instance, Kerner [2009] and Gartner et al. [2001] define congested traffic as the state in which the average speed is lower than the minimum average speed that is still possible in free flow. The transition from free to congested is characterized by an abrupt decrease in speed. At this point, the slope of the radius vector gives the minimum speed in free flow.

So far the theoretical considerations related to the fundamental diagram. The three curves in an empirical fundamental diagram depend on several conditions related to the location, topol-

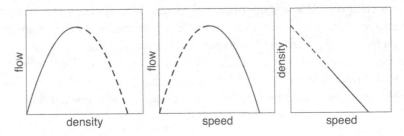

**Figure 2.1:** Theoretical fundamental diagram (dashed parts of the plots represent the congested regime; the transition points in each curve are the critical density (density-flow plot) and the critical speed (speed-flow and speed-density plots).

ogy, and purpose of the road section for which the fundamental diagram is being drawn, and may differ from the theoretical curve. For example, the fundamental diagram for an overtaking lane differs from that of a lane at its right; fundamental diagrams also differ under different weather conditions, for two-lane roads with opposing traffic; and so on. Further, the shape of the actual fundamental diagram depends on the length of the time interval over which the data is aggregated. It is desirable to avoid large aggregation intervals, particularly in the region of partially constrained and constrained flow, because it is easier to see the impact of individual slow vehicles and the stochastic element of the traffic flow. Finally, empirical studies must account for variability in the driver population. For instance, recreational routes may attract non familiar drivers, as well as regular drivers, and/or may have different populations of drivers on weekdays and weekends. Calibration thus involves observation and comparison of all possible situations/mixture of populations.

Because speed and flow are easier to measure, their relationship is often the best candidate for calibration studies based on empirical determination of an fundamental diagram. Chapter 2 in Roess et al. [2004] describes several calibration studies. Here, we focus on a speed-flow curve because freeways or other multi-lane facilities are best characterized by them. For sake of illustration only, we use a one-lane circular road of 5 km implemented by us in a microscopic simulator,[2] in which 180 vehicles travel continuously for 30,000 time steps (seconds), with speeds that can reach up to a maximum of 13.90 m/s (50 km/h). After all vehicles have entered the road, one is randomly chosen and its speed is reduced to the half for a time period that varies between 3 and 30 time steps. After this time, its speed is again reduced to the half for another random time period, until the speed eventually reaches a minimum value (which is close to but not zero). After this, the speed of this vehicle increases until reaching the maximum value. This way, a queue is formed behind this vehicle, dissipating after some time. This is done in order to simulate a situation of breakdown in the movement of the vehicles. The resulting plot appears in Figure 2.2. Each point is a pair $(\bar{v}, q)$ where $\bar{v}$ is the average speed of vehicles passing a cross

[2]We have used SUMO, also mentioned in Section 6.1.2.

section of the circular road (located opposite to the starting position). $q$ is the number of vehicles passing the cross section (over a period of 60 s). Points belonging to the congested regime were fitted using quadratic regression. In the uncongested regime, the fit is less clear (therefore we omit it here). However, we remark that the typical empirical speed-flow diagram has a discontinuity in the vicinity of the capacity of the road. This means that after a breakdown (abrupt decrease in average speed), the capacity may be lower than the capacity under normal flow.

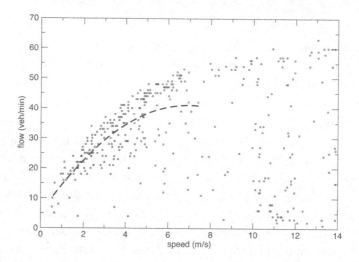

**Figure 2.2:** Example of speed-flow fundamental diagram (the dashed fitted curve represents the congested regime).

## 2.4   TRAFFIC CONTROL MEASURES

Although many traffic control policies exist, in this text we concentrate on traffic signal controllers (henceforth, traffic signals or simply signals) as a way to control the traffic by means of communicating simple messages to the road users so that safe and efficient states are achieved. Technically, this is called a signalized intersection (in this text, with some abuse of notation, the terms intersections, crossing, junction, traffic signal, and traffic light are used interchangeably).

Traffic signals can vary from hard-wired logic to computerized control, either centralized or not, as discussed later. Traffic signal controllers are connected to display devices (commonly, lights) that inform the road users when they may proceed with their movements, and when they must stop.

Increasingly, traffic signal controllers combine hardware and software in order to implement a signal timing that ensures that signal indications operate continuously, consistently, and efficiently. One important issue is that no conflicting movements receive green indication at the same time, unless other measures of priority are implemented. Traffic signals assign right of way

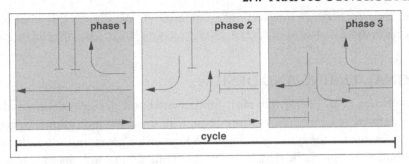

**Figure 2.3:** Signal plan with three phases.

to specific movements, thus eliminating or at least reducing the conflicts between some of these movements.

The U.S. Federal Highway Administration (FHA) publishes the Manual on Uniform Traffic Control Devices (MUTCD) for streets and highways, which serves as minimum standard for traffic signal based control in the U.S. According to the MUTCD, some advantages of using traffic signal control are: (i) to provide orderly movement of traffic; (ii) to increase the capacity (therefore the level of service) of an interaction (if the signal timing satisfies traffic demands); (iii) to reduce collisions; (iv) to coordinate in order to provide nearly continuous movement at a given speed along a given route (under given conditions); and (v) to interrupt heavy traffic to permit other kinds of traffic (e.g., pedestrians) to cross. However, poorly designed traffic signals can reduce these benefits. Moreover, some disadvantages of installing traffic signals are: (i) excessive delays; (ii) disobedience of the signal indications; (iii) avoidance of routes containing signals, thus leading to use of less adequate alternatives; and (iv) increase in rear collisions.

Traffic signals can operate in several ways. These are classified according to various dimensions. The first dimension regards the current traffic state; here the common modes of operation are pretimed and actuated. In pretimed operation, a sequence of signal indications, as well as their timings, is fixed and preset. It is repeated in each signal cycle constituting a plan that is designed to deal with a traffic volume based on historical data. In the actuated mode, the sequence and timing of some or all of the indications may change on a periodic basis, in response to a detected demand of road users. In order to work in actuated mode, controllers must be connected to detectors that provide information on, e.g., presence of pedestrians and vehicles, volume of vehicles, and other events.

The second dimension that determines the way a traffic signal operates relates to whether or not it operates in a coordinated (or synchronized) way. This requires a master controller or any other way of implementing the coordination among a set of individual signal controllers. In practice, this means that these signal controllers are coordinated or synchronized so that a progressive movement of vehicles is achieved. A common example is the implementation of a "green wave" over consecutive intersections in an arterial. We discuss this kind of operation in details in Section 2.4.2.

## 2.4.1   SIGNAL TIMING AND DESIGN

According to Roess et al. [2004], the most fundamental unit in signal design and timing is the cycle. This is defined as a complete rotation through all the green indications. In general, all legal vehicular movements (as for example in Figure 2.3) receive a green indication during each cycle.

A method to design traffic signals, including signal timing and phases, can be found in the HCM. Chapters 21 and 22 in Roess et al. [2004] also discuss this method. Here we give general ideas and guidelines underlying it.

The first aspect to be considered is an appropriate phase plan. In this task, a very important issue is the treatment of left turns (if allowed in the first place). It must be decided whether left turns are to be handled as opposing through flow, as protected movements (i.e., the opposing through movements are stopped), or as a combination of the two previous. Ultimately, the decision must consider both safety and the fact that protected movements require additional phases and contribute to lost time in the total cycle time. The HCM and other manuals bring guidelines for determining the appropriate phasing scheme.

Phase plans can be presented as phase and ring diagrams, in which arrows indicate the allowed movements in each phase. Figure 2.3 illustrates a phase diagram. Normally, solid lines indicate movements without opposition; dashed lines indicate opposed left or right turn; pedestrian movements can be shown as dotted or solid lines with double arrowheads.

As mentioned, the design of an intersection signalization can follow guidelines such as those proposed in the HCM. Fine tunings that are necessary can be made by means of traffic simulators (see Chapter 6 for a brief introduction of some of these tools).

Once a phasing plan is established, the next step is to assign appropriate timings for each signal phase in order to accommodate the demand. An important concept here is the one of split, i.e., the portion of green indication that each phase receives. It must be noted that although the split constitutes mostly of green indication, there is also the yellow indication and, some manuals also recommend the use of an all-red (clearance interval). In cases in which the legislation is such that it is legal to enter the intersection on yellow light, the all-red must provide sufficient time for all vehicles to cross the intersection before conflicting movements are given green indication.

The cycle length is the total time to complete one sequence of all movements around an intersection. To estimate the cycle length that will later be split into the various phases, an important step is the determination of the volume of the critical lane for each phase, as this controls the required length. However, this is not a trivial task given that the nominal volumes cannot be simply compared. For instance, trucks and buses require more time, as also traffic on right and left turns (with or without pedestrian movements allowed), and traffic on upgrade approaches. Thus, the nominal volume of the demand needs to be converted to equivalent through vehicle units (tvus), given in veh/h. Chapter 18 in Roess et al. [2004], for instance, gives formulas and tables to compute this equivalence, as well as the cycle length for a target tvus considering all critical lanes.

Typically, cycle lengths vary in the range 60–150 s [Bonneson et al., 2009]. However, there can be large deviations from these numbers. After the cycle length is determined, the effective green time is divided among the phases, proportionally to the critical lane volumes.

The aforementioned determination of phases and computation of cycle length and split characterizes a pretimed signal controller. We remind that in this case the cycle length, phase sequence, and timing of each interval are constant and follow a predefined plan. Of course, several signal plans (in which both sequence and intervals are different) can be determined to be used at different times of the day or days of the week. However, the use of these plans is predetermined.

A different situation occurs when it comes to actuated control, as this uses information about the current demand in order to modify the intervals and/or phase sequence. This information comes from some detection devices (see Chapter 5 for more details). Commonly, an actuated controller is programmed to alter the green time indication for each phase (and hence the cycle length) and/or the sequencing of the phases, so that the current demand is served the best way possible.

The benefit of an actuated signal timing varying on a cycle-by-cycle basis is significant. For instance, if a pretimed operation is designed to serve 30 vehicles per cycle in a given phase, but in a given cycle only 10 arrive, as the controllers interpret the detector signal as no demand, it changes the switches and interrupts the current phase. Conversely, if the demand is above 30 vehicles per cycle, the controller extends the cycle length. To avoid starvation and other undesirable issues, actuated control must observe: (i) a minimum and a maximum green time indication per phase; (ii) a way to determine what happens if there is no demand; (iii) fixed time for yellow and all-red intervals; (iv) adaptation in the pedestrian intervals; and (v) time to allow vehicles to travel from the detector to the stop line.

There are three types of actuated control. The full actuated requires that all lanes of all approaches have detectors. This type can accommodate optional phases, i.e., not only the phase sequence may change but the entire phasing itself. The semi-actuated type is employed where a major arterial or collector street intersects a small local street, in which detectors are installed. It receives green indication only when a vehicle is present. The interruption can be programmed in a way to cause less disruption in the flow of the arterial. The third type of actuated control is similar to the full one but it considers variable minimum green time, provided there are detectors capable of assessing the number of queued vehicles.

The disadvantages of actuated control are basically of two natures. First, there are additional costs involved, both for installing the detectors and for their maintenance. The second factor is more severe if progressive systems (next sections) are in use. Since this kind of system requires all signals in the progression to operate with equal cycle length, these cannot vary on a cycle-by-cycle basis.

## 2.4.2  PROGRESSIVE SYSTEMS

The goal of progressive systems (also called synchronized, coordinated systems) is to synchronize traffic signals along an arterial in order to allow a platoon of vehicles, traveling at a given speed, to cross the arterial without stopping at red lights. Thus, if appropriate signal plans are selected to run at adjacent traffic signals, a "green wave" is built, so that vehicles do not have to stop at intersections. The simplest way to implement a coordinated system is to compute synchronized signal plans for fixed times of the day such as morning and afternoon rush hours.

Progression imposes some restrictions on the phasing and timing of the signals that are to be synchronized. It was already mentioned that cycle lengths must be equal. This is required to ensure that the beginning of green occurs at the same time relatively to the beginning of the green indication at the upstream and downstream intersections. Exceptions are made in which the cycle length is a multiple of the basic lengths (rarely more than double).

The other restriction refers to the so-called offset and can be better explained by means of a time-space diagram, which is a plot of the traffic signal indications for two or more intersections, along time. Such plots convey the idea of vehicles' trajectories as a function of time. Figure 2.4 illustrates this with six intersections in an arterial (to keep the example simple, ideal progression is assumed). Let us assume that at time $t = 0$, the signal turns green at position $s = 0$. Vehicles there (as for example the car whose picture depicts its rear side in Figure 2.4) start or continue to move (in this case northbound), arriving at the sixth intersection (located at $s = 2000$) at $t = 140$. This car and all those entering the arterial at the first intersection from time 0–70 will be able to pass the whole arterial without stopping. The time interval between the start of the green time at two adjacent intersections is called the offset and depends not only on the distance between these intersections, but also on the average speed of the progression, which is normally indicated in a sign. There may be other definitions for the offset as, e.g., the distance relative to the initial intersection, but they are all variations of the basic concept just described.

Figure 2.4 also shows that there is a time window within a moving platoon of vehicles move through the series of intersections as signals turn green in order. This window is called the bandwidth of the progression and is determined by the split of the cycle length, but it may be smaller (and in practice it generally is) if conditions are not favorable. Trajectory lines in Figure 2.4 represent the speed of progression. Of course, if vehicles deviate from the recommended speed, this causes either a narrowing of the bandwidth (if the platoons travel at a lower speed), or the increase of stops (higher speed).

Also, this progression is a particular, simple, case in which the bandwidth is designed to use all the green time in one direction (assume that this direction has a much higher traffic volume).[3] Thus, one may expect vehicles traveling in the other direction to have to stop in some or all intersections. In fact, in this case, the progression is efficient only for vehicles traveling northbound.

---

[3]Readers can find a discussion on how to design green waves in more complicated cases in Roess et al. [2004].

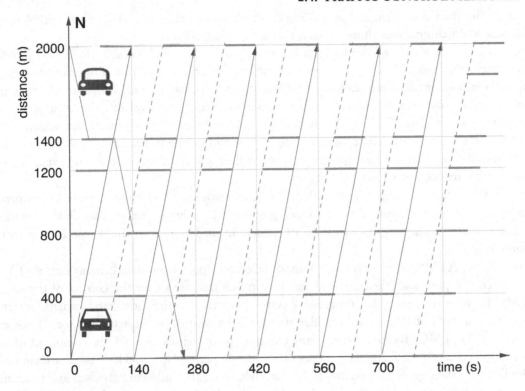

**Figure 2.4:** Time-space diagram of a progression in an arterial: equal bandwidths of 70 s.

The classical problem concerning synchronization systems is to find the optimal bandwidth for different cycle times and speeds. Popular solutions use hill climbing [Robertson, 1969] and mixed-integer linear programming [Morgan and Little, 1964].

A further problem is the consequence of such timing to vehicles traveling in the opposite direction as shown in Figure 2.4. Let us assume that a southbound vehicle (see car whose front is shown in the figure) leaves the sixth intersection at time $t = 0$ with the same speed as the bigger car traveling northbound. It is possible to see that this car has to stop at the fifth intersection ($s = 1400$) as well as at the third and first intersections. In each of these intersections, this car has to stop for a while, thus causing delays. In the end, it takes almost 280 s for the front vehicle to travel the 2000 m, while the rear car needs only 140 s for the same distance (in the opposite direction). This happens because the progression was originally designed for northbound vehicles. This is an extreme example as it sets the bandwidth equal to the green time for northbound vehicles. Although there are some algorithms that can optimize more than one bandwidth, the fact remains that, the more the constraints (e.g., in the form of more directions of progressions), the smaller the bandwidth, and depending on operation mode (e.g., real time) this problem may become a difficult one, resulting in bandwidths that are too short.

In fact, designing efficient offsets in two-way arterials is a classical optimization problem. Going even further, if one wants to optimize offsets in a grid or network of intersections, even if all are one-way, one does not scape the problem of network closures, i.e., the fact that determining some offsets following the aforementioned procedure fixes further offset(s) because of the constraints on shared links. Therefore, to avoid this problem, it is common practice to decompose the overall network and compute progression for non interdependent arterials. Further, the method just described ignores standing queues that might already be present when the first vehicle in the bandwidth arrives at the next intersection.

As the task of computing flexible and efficient progressive systems increase in complexity, more and more computerized tools are being employed. They are part of the ATMS (advanced transportation management system) effort, which, in turn, is part of the whole ITS effort (see Section 1.3).

In an ATMS, there is one or more so-called transportation management center (TMC), in which computers set the signals along the arterials that are under the control of the specific TMC in order to manage the progressive system remotely. In the most basic variant, detectors' information are not taken into consideration (i.e., the control is not actuated), or do not even arrive at the TMC. Rather, signal plans (running progressions or not) are computed off-line based on volume data for specific times of the day or days of the week. The advantage lies in the fact that the signals are updated from the TMC itself thus allowing changes and reaction to unexpected situations. Also, the TMC can keep a larger library of pretimed signal plans that can be employed on demand. Further, the TMC can keep statistics about failure as well as historical data. Increasingly, information from detectors is being used by the TMC. Even if this information is only used for monitoring purposes, it is useful as the TMC staff can manually either select the best plan or adjust the existing ones.

More recently, modern computerized systems have presented the tendency to full automation, also including virtual and on demand detectors based on camera images being sent to the TMC. Images are important tools for traffic monitoring because they may cover more observation points. An inductive loop detector can only deliver an instantaneous picture of a single point of measurement. Images quickly deliver a more general idea of the dynamics of the traffic flow; one that is readily processed by the TMC staff eyes, if not yet by an automated system.

Independently of being run off-line or at a TMC, there are computer programs to optimize the bandwidth. Next, we give an overview on them.

TRANSYT [Robertson, 1969, TRANSYT-7F, 1988] is a popular software for off-line optimization that generates optimal coordinated plans for fixed-time operation. Inputs are the geometry of the arterial, saturation flows, link travel times, turning rates at each intersection, demands (which are assumed to be constant), a set of pre-specified timings for the intervals, minimum green duration, and initial values for cycle time, splits and offsets. For given values of the latter three parameters, a model based on platoon dispersion is run, and a performance index is computed based on a combination of delays and total number of stops. It uses a hill-climbing optimization and its main drawback is that plans are computed for a static situation, based on historical data.

Examples of other tools from various generations and technological basis are: the Toronto and Washington experiences (see Section 24.8 in Roess et al., 2004); PASSER [Chang et al., 1988]; SCOOT (split, cycle, and offset optimization technique) by Hunt et al. [1981]; SCATS [Lowrie, 1982]; Prodyn [Henry et al., 1983]; OPAC [Gartner, 1983]; UTOPIA [Di Taranto, 1989]; and TUC (*Traffic-responsive Urban Traffic Control*) by Diakaki et al. [2002].

SCOOT and SCATS operations are adaptive and traffic responsive. They thus need real-time data. SCOOT uses data from detectors located at upstream end of the link. The main difference between SCATS and SCOOT is that SCATS is a hierarchical and distributed system. Data collection is local, based on detectors. For control purposes, an area is divided into smaller subsystems (1–10 intersections) that perform the control independently most of the time, i.e., appropriate cycle time and offsets are computed. Prodyn, OPAC, and UTOPIA are also adaptive programs in which control is not centralized. These do not consider explicit splits, offsets and cycles. In Prodyn, for instance, a decision is taken at each 5 s concerning whether to change phases or not. In a typical case, each intersection simulates all possible situations using detector information in adjacent areas. This information propagates from intersection to intersection with a decreasing weight. Both the relatively complex computation and the communication system can increase the cost of implementation. A more recent approach is the TUC, which was conceived for large scale networks. The coordination strategy consists of changing split, cycle, or offset (or all), as well as priorization of public transportation. Authors report results in two scenarios (small network, real world) simulating morning peaks. The performance was positive compared to a situation with fixed time synchronization.

Regarding the efficiency of progressive systems, it can be measured in various levels of abstraction. As a general measure, one seeks to optimize a weighted combination of stops and delays, a measure of the density (vehicles/unit of length) in the arterial or network, or travel time. However, particular characteristics must be observed. Number of stops and delay are acceptable measures for under-saturated arterials or networks. Here, queues are generally dissipated. In the over-saturated network, there is an excess of demand relative to the capacity, thus queues tend to expand over time, eventually blocking intersections. Control policies for over-saturated networks have the maximization of the throughput as primary objective.

There are limitations regarding the responsiveness of a traffic signal control. Changes in cycle length and offset cannot be implemented instantaneously and/or frequently. Rather, the transition may take some time. In general, it is not feasible to implement many different coordination patterns within a short period of time. Therefore, the current research trends focus on real-time adjustments that are performed carefully and efficiently, that can auto calibrate (such as SCATS), and that can act in a predictive way [de Oliveira and Camponogara, 2010].

In summary, progressive systems are valuable in some situations. However, the method-ology for creating a signal-timing makes a lot of simplifying assumptions, including a nominal volume of vehicles that reflects a "typical" condition. However, this is normally not the case. In traffic networks without well-defined traffic flow patterns like for instance morning and afternoon peaks, this approach may not be effective. This is clearly the case in large cities where business districts are no longer located exclusively downtown. Rather, there are several locations that serve as attractors for traffic so that no clear patterns exist. Also, in some cities, "secondary" streets have become as important as traditional arterials due to the saturation of these. Traffic patterns can also be affected by accidents, floods, snow, etc. Besides the already mentioned issue of cost and maintenance, as well as restrictions imposed in the design of the signal timings, a further issue is that it is difficult to optimize several intersecting arterials simultaneously in a grid-like network using real time data and in congested regimes. These issues show that simple off-line (or even on-line) optimization of the synchronization in *arterials* alone may not be able to cope with changing traffic patterns. Thus, more flexible and robust approaches are necessary. We discuss some recent approaches in Chapter 7.

## 2.5  TO KNOW MORE

Regarding traffic flow theories and models, the reader may consult the classical texts of Leutzbach [1988] and Gerlough and Huber [1975] (as well as its more recent revision at http://www.fhwa. dot.gov/publications/research/operations/tft/). Treiber and Kesting discuss diagrams based on cross sectional data and the fundamental diagram at length in Chapters 3 and 4 of Treiber and Kesting [2013]. Chapters 2, 6, 8, and 9 in Gartner et al. [2001] discuss traffic streams char-acteristics, traffic flow parameters, and traffic flow at unsignalized and signalized intersections respectively. For deeper discussions about traffic flow models and their ability to explain the phe-nomenon of traffic breakdown and other spatio-temporal patterns, we refer the reader to Ger-lough and Huber [1975], Helbing [2001], and Kerner [2009]. Readers interested in examples of fitting fundamental diagrams to observed field data are referred to Treiber and Kesting [2013] and Chapter 2 at http://www.fhwa.dot.gov/publications/research/operations/tft/.

Besides the mentioned High Capacity Manual (whose latest edition is from 2010), a pub-lication of the U.S. department of transportation [Bonneson et al., 2009] brings guidelines to design of traffic signal timings, as well as discuss progressive systems. Chapter 8 in Gordon and Tighe [2005] has a detailed description about the functioning of a TMC. For other forms of traf-fic control such as ramp metering, variable message signs, variable speed limits, high occupancy lanes, and road pricing (tolls) see Gordon and Tighe [2005], Sussman [2000], Roess et al. [2004], and Papageorgiou [2003]. Specifically, a review on road pricing (as regulator of excess of demand and hence controller of traffic congestion) can be found in Tsekeris and Voß [2009]. Classical works regarding this topic are Arnott and Small [1994], and Dial [1999, 2000].

# CHAPTER 3

# Elements of Demand

As a counterpart to the transportation infrastructure (discussed in the previous chapter), the present chapter introduces concepts related to demand. Contrarily to infrastructure, which is relatively fixed, the demand that uses such infrastructure changes frequently as it describes the behavior of users and their relevant choices. Traveling is no end in itself; rather, it is cause by people who need to perform activities at different geographic locations. Consequently, demand is highly differentiated by time of the day, day of the week, purpose, type of cargo, importance of duration of traveling, etc. (see Ortúzar and Willumsen [2001]). As mentioned in Chapter 1, although a relevant area, we do not consider freight transportation systems in this book. Some pointers to the literature covering this area are given in Section 3.4.

In Section 3.1 we focus on the two main approaches to modeling travel demand: (i) traditional trip-based and (ii) more recently proposed activity-based models. This constitutes the kernel of this chapter. As the central structure in which travel demand is represented is the so-called OD matrix, before discussing trip-based and activity-based modeling, we explain OD matrices. Section 3.2 describes discrete choice modeling. Section 3.3 then shortly discusses some questions related to travel demand management.

## 3.1 DEMAND MODELING

### 3.1.1 REPRESENTATION OF DEMAND

A central aspect of demand modeling is how demand is represented. Traditionally, this happens in a so-called "origin-destination-matrix" (OD matrix), a table in which each entry represents the number of users who want to travel from one location ("origin") to another ("destination"). For practical purposes, travelers' potential origin or destination locations are aggregated into transportation zones or districts. These zones may be defined in several ways based on statistical information gained from surveys, geo-referenced data collection, urban statistics, etc. A principle for determining these zones is that the variations of socio-economic characteristics inside the zones are as small as possible, whereas between zones the variations shall be as large as possible. Each of the zones is represented by its centroid. Its position is connected to the underlying road network and serves as an index for rows or columns of the table. Entires in the table are then the number of trips from the origin location (rows in the matrix) to a destination location (columns in the matrix). Figure 3.1 illustrates this.

Thus, an OD matrix is an aggregated representation of travel demand. Apart from the spatial differentiation, OD matrices may be segmented in other ways. For example, from a temporal

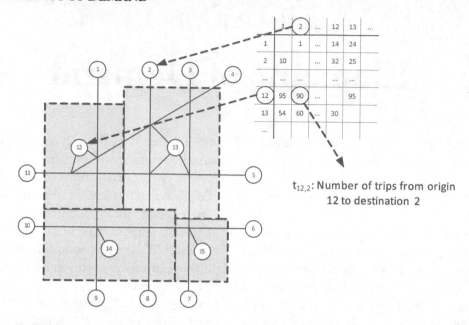

**Figure 3.1:** Basic concept of an OD matrix as a representation of aggregated travel demand (adapted from Barceló [2010], p. 7). Numbers represent different zones. In particular, the area of interest is divided into 4 zones (12, 13, 14, and 15), each represented by its centroid. Zones that are exogenous of the area of interest are marked from 1–11. Each zone is assigned a row/column in the OD matrix for representing how many trips originate/end in a given zone.

point of view, they may represent demands related to an ordinary working day, to specific peak hours, etc. Further, OD matrices can represent origin-destination relations for particular modes of transportation, such as OD matrices for individual vehicular traffic, for public transportation, or also for different transportation purposes.

## 3.1.2   TRIP-BASED DEMAND MODELING

Trip-based approaches to demand modeling are the core to the classical "four-stage model"[1] of traffic modeling [Ortúzar and Willumsen, 2001]. Three of these stages refer to demand modeling: 1. Trip Generation; 2. Trip Distribution; and 3. Modal Split. The fourth one—Assignment—connects demand to supply. These steps are depicted in Figure 3.2.

Next, we shortly address steps 1–3, each resulting in an OD matrix. Step 4, trip assignment, actually refers to locating trips to links in the network, that means, determining which share of users travels on which link. Due to the importance of trip assignment, the next chapter is completely dedicated to the fourth stage.

---

[1]In the literature, one can find different terms denoting these four subsequent elements: "four phases," "four steps," etc.

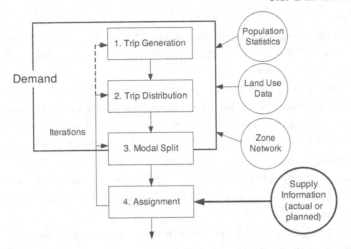

**Figure 3.2:** Schematic overview of the classical four-stage model of transportation modeling (adapted from Ortúzar and Willumsen [2001], p. 23).

## Trip Generation

This stage answers questions such as how many trips are generated in a transportation zone, and how many are attracted (end) in a zone. Thus, it produces the number $T_i$ of trips originating in each zone $i$ and number $X_j$ of trips ending in each zone $j$.

There are two main approaches to trip generation, which are based on statistical regression models connecting attributes of a zone with the number of trips produced in or attracted to a zone:

- On the aggregate level (meaning values for the full zone): the number of trips is assumed to be a function of zonal properties such as number of employees, inhabitants, roofed area associated with particular services, etc. For example, a residential area with a given number of houses, mean household size, and number of available cars produces some number of outgoing trips in the morning.

- On the disaggregate level (which means that the number of trips is calculated for each household and then aggregated): the number of trips depends on individual household characteristics such as income, motorization, household structure, accessibility, etc. This requires an additional aggregation step for acquiring zonal values, but has the advantage that the generation of trips happens on the level of the actual decision makers.

For consistency, the overall number of attracted and of generated trips must be the same: $\sum_i T_i = \sum_j X_j$. There is a number of techniques to assure this. For more details on using statistical methods at different levels of complexity, see Ortúzar and Willumsen [2001] and Oppenheim [1995].

## Trip Distribution

In the trip distribution stage, produced and attracted trips are connected, and trip numbers are distributed to cells of an OD matrix. Therefore, this steps produces number of trips $T_{ij}$ originating in zone $i$ and terminating in zone $j$.

Similarly to trip generation, there are two basic approaches: (1) disaggregated using explicit destination choice depending on individual characteristics or (2) aggregated starting from the summation of trips that should start or end in each zone, and then distributing these trips. For instance, using Figure 3.1, if it is known that, say, 1,000 trips originate in zone 12, then these must be distributed among zones that consume them, resulting in the numbers shown in that figure. Typically, there is no feedback between trip generation and trip distribution in the standard four-stage model. That means, trip generation is not affected by the attributes of travel destination, available travel modes, or travel routes.

Typical disaggregated trip distribution models are discrete destination choice models. Hereby, a utility function is estimated based on survey data in which attributes of households and origin zone are used to determine the most probable destination zone. In Section 3.2 we will give more details about this general approach that finds application for many subproblems in travel demand modeling.

Typical methods to implement aggregated trip distribution are based on analogies with other methods describing similar phenomena. One of them is the "gravity" model which is based on the observation that flows between two districts depend on a "resistance." Such a resistance function $f$ can contain factors describing (most often) costs, travel time, or distance for a trip from origin zone $i$ to destination zone $j$. This travel resistance can be calibrated based on trip frequencies obtained from surveys or statistics. A general formulation of the gravity model for the number of trips from $i$ to $j$ $T_{ij}$ can be given as in Eq. 3.1, where $Q_i$ is some measure of productivity of the zone $i$, usually the number of trips starting in $Q$; $A_j$ is some measure for the attractiveness of destination $j$, usually equal to the number of trips ending in $j$; $f(c_{ij})$ is the resistance function depending on travel costs $c_{ij}$; $\mu$ serves as a weight for the combination of $i$ and $j$.

$$T_{ij} = \mu Q_i A_j f(c_{ij}) \tag{3.1}$$

For appropriate entries in an OD matrix, the following constraints need to be fulfilled:

$$\begin{aligned} \sum_j T_{ij} = T_i \quad &\forall i \\ \sum_i T_{ij} = X_j \quad &\forall j \end{aligned} \tag{3.2}$$

This means that the trips going from a cell $i$ to any other cell must sum up to the overall sum, and similarly, all trips incoming to a cell $j$ from different directions must sum up to the overall sum. This is not guaranteed just by calculating a gravity model. Rather, it is achieved in a separate step: A way is to enter the values generated by the gravity model into the OD matrix, sum up values in the cells and rescale rows, sum up values in the rows and rescale columns, and so on until convergence has been reached.

More details on the gravity model can be found in Ortúzar and Willumsen [2001] or Cascetta [2009]. The gravity model belongs to the area of so-called "spatial interaction modeling." Alternative approaches for connecting possible origin zones and relevant destination zones to trips are available. A review can be found in de Vries et al. [2001].

## Modal Split

In the third stage of classical demand modeling it is determined which trips are covered by which transportation mode. The result of this step is a set of OD matrices, one for each mode: each cell contains the number of trips $T_{ijm}$ originating in zone $i$ and terminating in zone $j$, using a mode $m$. Hence, the overall number of trips between two zones is "split" into parts for each mode.

The consistency condition $\sum_m T_{ijm} = T_{ij}$ $\forall_{i,j}$ must hold, stating that the sum over all modes must equal the overall number of trips in that OD matrix. This clearly depends on characteristics of the trip maker (e.g., car availability), characteristics of the trip (e.g., trip purpose) and of particularities of the transportation mode, such as the cost when taking public transportation, or the load that can be transported, etc. There are different approaches for this: (a) "trip-end modal split models" that are based on demographic characteristics of the producing or attracting zone; (b) "trip interchange modal split models" that distribute the number of trips between two zones as a function of costs or other characteristics for each mode; and (c) combination of destination and mode choice. As we focus mainly on vehicular-traffic here, we skip the details of these methods. An elaborate description can be found in Ortúzar and Willumsen [2001].

## Advantages and limitations of the four-stage model

Although the fourth step of the four-stage model will only appears in the next chapter, here we anticipate some criticisms as well as advances of the four-stage model, which are independent of the last stage.

The four-stage model has been criticized for several aspects. The main ones are as follows. First, there is no consistent modeling of the travelers' decision making, as the steps are decoupled. This means, for example, that while the trip generation module knows about households, income, daily plans, etc., other modules only know the starting and ending point of a trip. Second, it is appropriate mostly for trips that originate at homes, without regarding subsequent trips of the same person or their contexts. If new technologies such as ATIS and other guidance systems that are discussed in Chapter 9 affect choices of different segments of the population differently, then this approach cannot consider this. Third, the treatment of demand based on the four-stage model is highly aggregated. Finally, it has problems at capturing the impact of dynamic measures for demand management (see Section 3.3) such as toll or congestion pricing, etc. The level of abstraction is too high for reproducing realistic elastic reactions.

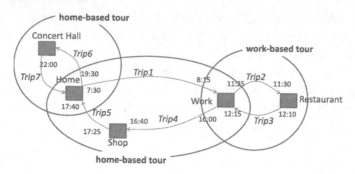

**Figure 3.3:** A typical daily plan containing a number of trips which connect different activities.

Nevertheless, the four-stage model is still highly relevant in the traffic planning community, in which reproducing disaggregated and context-dependent decisions are not as relevant as in traffic flow simulation. Also, it is useful when the time scale in which the system is modeled is high. In Chapter 6, traffic planning and traffic simulation are shortly contrasted. Integration of intelligent techniques to improve the realism of such models is limited, yet these forms of demand modeling can be sufficient for providing input to testbeds for intelligent methods supporting route choice, etc.

### 3.1.3   ACTIVITY-BASED DEMAND MODELING

Activity-based demand modeling is the approach with most relation and need for intelligent solutions as it directly aims at reproducing human decision making in daily life. Whereas trip-based demand models consider each trip from one zone to another in isolation (often with the assumption that trips are starting or ending in the home zone), the basic starting point of activity-based demand models is the idea that trips are not independent from each other, but organized in chains of trips. Figure 3.3 illustrates a case in which trips depend on each other. Travel demand must be derived from activities that occur at different locations. Time plays an important role beyond just a cost-factor for decision making. Individuals also decide about when to travel. Thus, activity-based models capture the necessary context of traveling, so that—in contrast to four-stage model—they are particularly apt for testing potential effects of different, also dynamic strategies for influencing decision making on departure times or travel mode. Activity-based demand modeling is not just about connecting different activity-locations. Rather, it can deal with issues such as the fact that a delay in a trip may influence whether another trip even occurs at all.

Early approaches of activity-based demand modeling were based on spatio-temporal constraints[2] as a framework for combining temporal and spatial processes [Miller, 1991]. It provides a language to integrate and reason about spatial and temporal constraints and thus allows rep-

---

[2]This framework is also known under the term Time Geography originally introduced by T. Hägerstrand analyzing migration processes [Lenntorp, 1999].

resenting available time periods for traveling using so called "space-time prisms." For each zone the earliest arrival time as well as the latest departure time for a given transportation mode is determined. This information is then represented in a prism expressing spatio-temporal constraints providing the frame for traveling, possible destinations, modes, etc. Timmermans et al. [2002] relate this research on space-time behavior to various activity-based demand models.

More recent approaches to activity-based demand modeling use AI approaches for reproducing the decision making concerning daily activity plans. More elaborated models that formulate how activities can be sequenced or combined, can be found in activity scheduling or activity switching models.

In principle there are two types of approaches for elaborating these activity scheduling: (i) direct approaches in which the decisions of travelers for each element of a daily plan are emulated or (ii) iterative approaches in which activity schedules are optimized assuming that the best one is the most realistic plan. In the following, we shortly discuss these two approaches.

Initially, activity plans are given as plan skeletons. These are basis for typical plans for a particular category of travelers (working person, school child, etc). Then, they are enhanced with other activities. Primary activities are assumed to happen at given locations (for like home and work); only secondary activities such as shopping or leisure ones are subject to location choice. Hence, activities "only" have to be configured, scheduled, and then connected by a corresponding trip. A rather simple example can be found in Rindsfüser and Klügl [2005]. In ALBATROSS [Arentze and Timmermans, 2004], a rule-based approach based on decision trees is used. For every category of individuals there is a set of fixed primary activities and a set of flexible, secondary ones (e.g., daily shopping). In contrast to the skeleton-based approach, the reasoning of the simulated users happens in a given sequence: first, members of a simulated household start selecting the transportation mode associated to fixed activities. After, they add flexible activities and decide on their duration, time, etc. Decision of all household members are done in an interleaved way. These decisions add dynamic constraints restricting the subsequent ones. The result is then a full daily plan for each individual. Each decision is based on a decision tree that is learned from surveys with activity diaries filled by human subjects during several days. In a later research, decision trees were replace by Bayesian networks [Arentze and Timmermans, 2008].

A second, iterative approach is used in the MATSim-T system [Balmer et al., 2009]. Here, every member of a synthetic population possesses a set containing several fully elaborated daily plans. Fully elaborated means here that a plan also includes the routes between the activities' locations. Daily plans are optimized using evolutionary approaches, in a way that assuming that the "best" plan corresponds to the most realistic one. Appropriate mutation operators may modify durations of activities, departure times, transportation modes, routing etc. In each simulation, a traveler selects one plan out of her set of plans. The plans of the complete population are simulated and then evaluated using utility functions that consider not only the utility of all activities measured in a given value, but also their duration, as well as dis-utilities for eventual late arrival at the next activity in the chain.

These are two examples of activity-based demand modeling; more can be found in, for instance, Timmermans [2005]. An earlier review can be found in Axhausen and Gärling [1992], while discussions from different points of view are given in Bhat and Koppelman [1999]. In summary, activity-based demand models use complex approaches that aim at reproducing human decision making about daily plans, including trips to connect activities. They directly use the fact that traveling is a derived activity, and not an aim of its own. Due to the intended high level of detail, many assumptions have to be made, either directly on how the decisions are make, or on how humans evaluate the different elements of a daily plan.

In general, activity-based demand modeling puts a high demand on data for validation. It requires not just observations of traffic load, but also information on activities that are gathered in expensive surveys. Many persons have to be found, who do full recordings of their locations and activities for at least one complete day. With modern GPS and smart phone technology, the acquisition of such data becomes more and more feasible. We will tackle this in more detail in Chapter 8.

## 3.2   DISCRETE CHOICE MODELING FOR TRAVEL DEMAND

An activity-based demand model embeds a variety of choices, ranging from selection of secondary locations, to departure time choice, mode choice, route choice or even choice of lifestyle that then determines the daily plan skeleton. Pinjari et al. [2011] speak of a "choice continuum." So-called discrete choice models form a way to build models capturing these choices in isolation or in a combined way. They can support activity-based as well as trip-based demand modeling.

Discrete choice modeling promises to produce more realistic and powerful models than the classical four-stage model, as it is directly based on the observed choices of individual travelers. In principle, discrete choice modeling can be applied in all cases in which travelers need to select one out of a number of discrete options. Such discrete options can be modes or transportation, routes, or simply a binary decision such as between staying on a route or taking a detour. Hence, these models can also be used within an activity-based demand modeling framework. He just give a brief introduction here. More information and examples can be found in Ben-Akiva and Lerman [1985], Ben-Akiva and Bierlaire [1999] and Train [2003].

In general, the basic assumption of discrete choice modeling is that "the probability of individuals choosing a given option is a function of their socio-economic characteristics and the relative attractiveness of the option" [Ortúzar and Willumsen, 2001, p. 220]. This attractiveness is represented as utility, which the individual traveler tries to optimize. Yet, the utility that an individual assigns to an option is not fully known to the modeler; rather, only a part of it, the "observable utility" is known. There is a number of assumptions that need to be taken to determine the particular structure of the model. The first assumption concerns the real utility function $U_{i,o}$ with which a user $i$ evaluates an option $o$. The real utility function has two parts: the observable utility $V_{i,o}$ and the error term $\epsilon_{i,o}$: $U_{i,o} = V_{i,o} + \epsilon_{i,o}$. The observable part is based on attributes of the option and the individual. The error term collects all unobserved factors that may affect the users choice, but are not measurable or accessible. A second assumption is that the error associated with all those unobservable factors is a random variable with a particular probability distribution that determines the particular type of the discrete choice model. Another assumption is about the structure of the observable utility function: Often, a basic structure is used, like $V_{i,o} = \beta_{i,0} + \beta_{i,1}X_{i,1} + \beta_{i,2}X_{i,2} + ...$, where each $X_{i,n}$ describes measurable attributes of the option or of the decision maker. The $\beta$'s are the weights of the different attributes. Their values are set in an explicit model estimation step based on empirical data: In surveys, human subjects are asked to evaluate and select different options. Model estimation of a behavioral model means that the weights of the observable utility function are optimized with the goal that the likelihood that the decisions of a simulated population match the recorded choices of the survey is maximal. Once the weights are estimated, they can be used for simulating choice behavior or serve as the basis for interpretation of which attributes of the different options are relevant for the surveyed population, for that particular decision.

A very simple example may illustrate this: A typical discrete decision problem is the choice whether to take the bus or the car to travel between two locations in a particular area. The initial steps of developing a discrete choice model are (i) to setup a hypothesis for a valuation function (in this case a weighted sum of travel time and travel cost) and (ii) to survey a sufficiently large share of the population of the study area, and capturing under which circumstances people would decide for which option. In model estimation, the values of the two weights $\beta_{time}$ and $\beta_{cost}$ are set so that the decisions of the virtual individuals are as similar as the decisions of the real individuals. When this is done, the discrete choice model is elaborated and can be used also in slightly different scenarios.

For predicting which option an individual might choose, the user's valuation of one option needs to be expressed in relation to the valuation of all other options, to be normalized, and transformed into a probability. The most prominent of such functions is the so called *logit* function. For instance, given two alternatives, the *logit* function in Eq. 3.3 gives the probability to choose alternative 1 with valuation $v_1$:

$$P_1 = \frac{exp(v_1)}{exp(v_1) + exp(v_2)} .$$ (3.3)

This approach has some restrictions. The most problematic one is that options must be independent, which is for example not the case if two alternative routes share edges. Yet, in summary, if appropriate data on real human choices is available, discrete choice models are very powerful and reliable tools combining strong hypothesis on what is important for decision makers, with statistical evidence.

## 3.3  TRAVEL DEMAND MANAGEMENT

With the objective of improving the overall traffic situation, but also to support the individual traveler, demand management focuses on the idea of tuning the time or the mode of a trip, or even adapt the activity context of the potential traveler. For instance, making communication network and data sharing infrastructure available for working at home influences travel demand. The overall goal is to avoid congestions and to better distribute traffic infrastructure usage in time and mode by adapting the demand.

Winters [1999] illustrates, from a user's point of view, how travel demand can be influenced in the future. This can be based on individualized information about the state of the traffic system, thus enabling the user to spontaneously adapt departure time, or even to cancel travel at all and deciding to work at home instead. Also, flexible car sharing has effect on travel demand. Early works (e.g., Wachs [1991]) have focused on behavioral change; meanwhile one can classify many approaches that provide information to the travelers as belonging to the area of transportation demand management.

## 3.4  TO KNOW MORE

This book is only concerned with transportation of persons, more precisely with vehicular traffic. Freight transportation and transportation logistics are similarly established topics that are highly relevant to the economy, environment and society. There are several sub-areas ranging from supply chain management and operation to last-mile delivery systems.

A survey of agent-based approaches in logistics and freight transportation modeling can be found in Davidsson et al. [2005]. Liedtke [2009] presents a model which on a high level of details based on commercial activity chains. Holmgren et al. [2012] integrate logistics simulation with the optimization of production schedules, while Joubert et al. [2010] integrate freight vehicles and private cars in an agent-based simulation. These are just a few examples, additional reviews can be found in Chow et al. [2010], Crainic and Laport [1997], and Zhou and Dai [2012].

# CHAPTER 4

# Traffic Assignment: Connecting Supply and Demand

In order to actually predict the load or flow on the road network, it is necessary to determine which parts of the infrastructure the travelers actually are willing to use. This means, the demand part of the four-stage model (see Chapter 3) needs to be assigned to the available supply (Chapter 2). Thus, the so-called traffic assignment (or, alternatively, route assignment or trip assignment) models the interaction between the traffic system and drivers decision making. The main challenge in modeling and controlling transportation systems is the difficulty (if not impossibility) of a system manager to directly control the behavior of the drivers in terms of their route choices.

In the classical four-stage model (see Figure 3.2), traffic assignment forms the last phase. Its output describes the state of the transportation system, which is a relevant input for evaluating the consequences of changes in the infrastructure. Moreover, for microscopic traffic flow simulation, information about routes chosen is relevant, either directly for representing the decision of a simulated road user at an intersection, or indirectly for determining turn probabilities at intersections. The information necessary for this is generated by the process of traffic assignment.

Basically, travelers select routes for connecting their individual origins with their individual destinations. In a traditional aggregated and static view, the trips in the OD matrix are assigned to links that are part of a route from an origin node to a destination node. User decisions and thus demand flows are influenced by the costs of the single edges forming the route, which depend on other travelers decisions about the links they use. More recent approaches deal with dynamic traffic assignment, thus focusing on the issue of time-dependency.

Traditionally, there are many traffic assignment models with approaches based on different assumptions. Cascetta [2009] gives a fundamental classification along the following dimensions.

- **Interaction:** The most important dimension is which assumptions are made about the interaction between demand and supply. The selection of a traveler has an effect on the traffic load and thus on the utility of a particular path for the other travelers. Thus, an important category of methods is composed by user equilibrium approaches, in which the distribution of trips is consistent with the costs they produce. We discuss these approaches in the next section. Alternative approaches are day-to-day (between-period) approaches, assuming that the system evolves over time converging to a consistent, equilibrium state. These algorithms are based on different assumptions about the information possessed by travelers and their path choices.

- **Supply factors**: There are two factors that are considered for the supply: whether the supply offers continuous services such as roads for vehicles, or only on a scheduled basis, such as public transportation. An essential dimension is whether there are assumptions on the feedback loop between load and travel-time, i.e., whether assignment is done in a un-congested network or a congested one.

- **Demand factors**: There are four dimensions to distinguish in this category: whether there is only one class of users, or multiple classes ("Demand Segmentation"); whether demand is fixed, or variable in reaction to load ("Demand Elasticity"); whether route choice is done completely pre-trip and does not change during evaluation, or whether route choice is done pre-trip as well as en-route ("Path Choice Behavior"); and finally, whether there is a deterministic or a stochastic route choice ("Path Choice Model").

In the following we will first discuss how routes can be computed and then give some examples for existing traffic assignment methods.

## 4.1  ROUTE COMPUTATION

The connection between origin and destination usually does not consists of a single link, but of a sequence of links passing through nodes, together forming a route. The most relevant set of algorithms for determining the relevant path(s) or routes are "shortest path search" algorithms. These algorithms belong to the standard material that is taught in computer science classes. Relevant algorithms here are the Dijkstra algorithm [Dijkstra, 1959] or variants of A*. Both are based on costs assigned to each link. Costs in traffic scenarios are "generalized costs" which contain travel time as the main component. Other relevant factors in such cost-function can be tolls, prices for public transportation, parking availability and cost, comfort, or the value of time depending on the mode of transportation. These cost functions may be individualized to mirror different weights for the different attributes, but also different perceptions and interpretations of costs. For example, the later may be a function of the traveler's wealth.

Also, modern navigation systems use these algorithms. In Chapter 8 we provide a deeper look into approaches for efficient routing for route guidance systems. In traffic assignment, variants of such algorithms are used to determine a set of alternative routes between origin and destination nodes, so that trips can be distributed with respect to complete routes and not merely based on individual links.

## 4.2    BASIC TRIP ASSIGNMENT

### 4.2.1    ASSIGNMENT ON UN-CONGESTED NETWORKS

The rationale regarding simple trip assignment is to assign all trips to the route with minimum cost, on the basis that these are the routes travelers would rationally select. No feedback loop is assumed between route selection and costs. That is as in Eq. 4.1, where $T_{ijm}$ is the given modal demand between $i$ and $j$ obtained from demand models. This procedure is referred as static "all-or-nothing" assignment.

$$\left. \begin{array}{ll} T_{ijmr^\star} = T_{ijm} & \text{for the minimum cost route} r^\star \\ T_{ijmr} = 0 & \text{for all other routes} \end{array} \right\} \quad \forall_{i,j,m} \qquad (4.1)$$

All-or-nothing assignment is not useful in congested systems as it assumes that the influence of the assignment of the trips to the route can be neglected. Additionally, not all travelers take the same route due to personal preferences and perceptions. Thus, a stochastic version may distribute a share of trips onto different routes depending on their utility. Before this can be done, a number of full alternative routes have to be generated. There is a number of algorithms for determining several good alternative routes, such as k-shortest paths [Yen, 1971], shortest path algorithms repeated with link costs plus some random noise, using different variants of the cost functions, or proportional stochastic methods, such as the algorithm of Dial [1971].

A basic problem of such assignment methods is that, if trips have a non negligible set of edges in common, the distribution of the trips onto the different routes may result in all (or many) trips passing through one (or few) link(s). This happens due to the assumption of trip independence.

### 4.2.2    ASSIGNMENT UNDER CONGESTION AND EQUILIBRIUM

As mentioned in Chapter 1, there are nonlinear interactions between travel behavior, route choice or routing, and congestion. In a situation in which every user has found a route with the least travel time or minimum cost, each has no incentive to change route as every other route would be more expensive. This situation constitutes a so-called user equilibrium. Such an equilibrium situation is persistent as long as the demand and the network do not change. This concept is also called Wardrop's first principle [Wardrop, 1952] or user equilibrium.

No congestion in the network basically means that the demand does not depend on travel volumes. In that case, simple assignment methods are sufficient. However, in presence of congestion the situation becomes significantly more complex. Link and destination travel costs are function of the demand. Yet, they may influence the demand for some or all OD pairs, as users might delay departure or change the destination location due to high costs. Thus, it is easy to see that there is a complex feedback relationship between travel demands at all four stages. Ideally, all phases of the four-stage model should be solved simultaneously to obtain levels of demands that are in equilibrium at the assignment stage. This has been realized by full agent-based traffic simulation models as for example MATSim which has been introduced in Section 3.1.3. As explained there, daily plans there are elaborated in such a fine way that they also contain the chosen routes for traveling between two activity locations. Traditionally, iterative procedures are used for assignment, assuming a given demand. In principle, they work according to the cycle depicted in Figure 4.1.

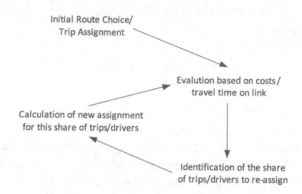

**Figure 4.1:** General scheme of iterative traffic assignment.

Mainly, three simple iterative methods are used: the first is to start from some initial values for the link costs and find routes (for each OD pair) with minimum costs. Trips are then assigned to these routes. New costs are computed and this cycle is repeated until there is no significant change in link or route volumes. However, there is no guarantee that such a stable situation will be reached, thus oscillations may occur as portions of the trips keep being assigned back and forth to the same set of links.

The second approach is to load the network incrementally, e.g., assign a given fraction (i.e., 10%, 20%, etc.) of the total demand (for each OD pair) at each time. Further fractions are then assigned based on the newly computed link costs. This procedure continues until 100% of the demand is assigned. There is also no guarantee that an equilibrium will be found, as this forms a straightforward, greedy approach without iterations.

In contrast to these two, the method of successive averages finds an equilibrium assignment (assuming infinite iterations): here the load is determined using a repeated all-or-nothing assignment, in which costs are determined anew by mean trip distributions over all previous iterations. Convergence may be very slow. Established traffic assignment tools (see Chapter 6) use multi-level algorithms that determine an equilibrium for existing route sets and then check whether there are new reasonable routes. There are algorithm that are faster than the method of successive averages. The Frank-Wolfe algorithm [Frank and Wolfe, 1956] is one of them but it is still not applicable for big networks. More recent algorithms (gradient projection and bush-based methods [Gentile, 2012]) are orders of magnitude than the method of successive averages.

This concept of such a static[1] equilibrium has been criticized for being based on several simplifying assumptions: (i) network characteristics are fixed and known to the users; (ii) similarly, OD flows are known; (iii) users know or perceive travel times throughout the network; (iv) users choose routes that minimize travel time or a cost function; and (v) travel times and volumes on links are considered to be constant and just dependent on the assigned load.

Despite these criticisms, this static equilibrium is a useful concept when the modeling horizon is long enough, as for instance for long-term planning. In this case, it is reasonable to assume that users do discover new routes with a lower cost. Further, in a long time horizon there is no reason to account for variations in travel times and volumes on links. Assumptions (i)–(iii) can be justified by on-board devices such as in-vehicle navigation systems, as discussed in Chapter 8.

Static assignment methods are based on a commonly agreed-upon formulation—in contrast to dynamic traffic assignment. Mathematical properties such as existence and uniqueness of equilibrium can be obtained relatively easily, and the computing time necessary to find the equilibrium solution (or a reasonable approximation) is, in general, low. Finally, static methods are robust to errors in the input data and require less data to calibrate.

As in the un-congested case, the method can be extended using stochastic assignment. The stochastic user equilibrium (SUE) [Daganzo and Sheffi, 1977] was introduced with the aim of relaxing the unrealistic assumption of users having perfect knowledge of travel costs, which underlies the classical Wardrop (user) equilibrium. SUE reproduces a particular distribution of travelers over a set of routes. Then, in an iterative procedure, the distribution is adapted by moving load from highly loaded routes to less loaded ones.

## 4.2.3 DYNAMIC TRAFFIC ASSIGNMENT

If the focus is on short-term modeling and/or at a fine-grained level, then equilibrium-based, static approaches are not suitable because they do not consider short time variations that occur in the traffic flow. As a matter of fact, an actual steady state practically does not exist in traffic networks, unless data is considered on a long time horizon. Thus, traffic assignment methods that consider the dynamic nature of flow are necessary. Such methods are subsumed under the umbrella of so-called dynamic traffic assignment (DTA). By implementing two modifications to static assignment, DTA is able to consider variation (over time) in travel times on links [Chiu et al., 2011]: The first modification is that travelers are assumed to know/anticipate *future* travel conditions along the journey between an OD pair. They aim at minimizing the travel time considering the condition of a link at the moment in which they will actually use it. This depends on when they arrive at the various links along a route. The second modification derives from the former: in DTA, since travelers who depart from an origin to a destination at different times will experience different travel times, the user equilibrium condition applies only to travelers who are assumed to depart at the same time between the same OD pair.

---

[1]This is named "static" as there is no time-dependency of load assumed.

The just-mentioned first modification requires the description of time-varying network traffic condition and finding least cost routes when link travel times change over time. Analytically, this can be accomplished by so-called time-dependent shortest path (TDSP) algorithms, which minimize the actual experienced travel time. Apart from these analytical methods, simulation-based approaches also exist, as discussed in Chapter 6.

The second modification is called disaggregated equilibrium since it establishes an equilibrium condition for each departure time, where these are discretized in assignment intervals (seconds to few minutes). To distinguish from the standard static equilibrium, it is called dynamic user equilibrium (DUE[2]). The solution of a DUE is not trivial: it requires finding a set of time-varying link and route volumes and travel times that satisfy the DUE condition for a given network and time-varying OD demand pattern. The difficulty arises from the fact that the route with least cost for each traveler depends on congestion levels throughout the journey, which in turn depend on the route choices of other travelers who have departed earlier, are departing at the same time, or will depart later, since all these trips may share links. Again, solutions must then be found through an iterative process, starting from some initial set of route choices, and gradually improving them. In practice, to avoid long iteration times, many approaches to DTA/DUE find an approximate equilibrium (e.g., epsilon-convergence, in which arbitrarily small changes in link costs are allowed).

Typically, this iterative process involves a phase in which routes are evaluated, i.e., the effects of all route choices are determined; a second phase where the routes with the lowest experienced travel times (for each OD pair and assignment interval) are determined; and a third phase where adjustments are made in the assignment coming out of the second phase. As in the static case, if a fraction of the routes change to others with a lower experienced travel time routes, then the assignment can be brought closer to equilibrium. The reason why normally just a fraction of the route choices are is that some routes can then become congested and this would no longer be an equilibrium since given travelers would be better off changing routes. This assumption of changes by only a given fraction of the route choices is important to avoid the appearance of travelers' overreaction.

## 4.2.4   FROM USER TO SYSTEM OPTIMUM

Given a traffic network represented by a directed graph (see Chapter 2), the assignment from the point of view of the user (user equilibrium, UE) can be analytically stated as an optimization problem: find all flows from each OD pair s.t. only paths with minimal costs have a nonzero flow assigned to them, which corresponds to Wardrop's first principle.

For a mathematical formulation of this problem, the reader is referred to Chapter 2 in Gawron [1998a], as well as to Ran and Boyce [1996] or Cascetta [2009].

Determining the system optimum means minimization of the total travel time of all travelers. According to Gawron [1998a], just as the user equilibrium problem can be formulated as an

---

[2]We follow the acronym as in Gawron [1998b] and Mahmassani and Herman [1984].

optimization problem with respect to a special cost function, the system optimum problem can be formulated as a user equilibrium problem with respect to marginal costs. Since rational behavior by the individuals leads to the user equilibrium, these marginal costs are the costs that should be enforced by a network operator in order to ensure that the rational behavior of the individuals leads to the system optimal traffic flow pattern. This is especially important in traffic management applications where the objective is to apply some control measures to get the traffic flows to the system optimal state.

Both the UE and the system optimum solutions can be numerically solved by the Frank-Wolfe algorithm [Frank and Wolfe, 1956], which can solve optimization problems where the objective function is convex and the constraints of the problem are linear. This algorithm converges to an exact solution. Moreover, it can be used to find the minimum cost flow with respect to the link costs by fixing an initial solution. Assumptions here are that links have no capacity limits. Thus, this problem is equivalent to solving a shortest path problem for each OD pair. These shortest path problems can be solved very efficiently, e.g., using Dijkstra's algorithm. However, a shortcoming of the Frank-Wolfe algorithm is its speed of convergence, which is too slow for most practical uses, even considering acceleration methods that have been proposed (see Buriol et al. [2011] or Gawron [1998a] for further references). In Gawron [1998a], it is mentioned that the theoretical convergence rate is arithmetic (although typical cases have better convergence rates).

Regarding uniqueness of the solution for both the UE and from the point of view of the system, one can only guarantee that there exist a unique solution for each link flow. However, this may not be true for path flows, because there may be different path flows yielding the same link flows. An example of such a situation is when two OD pairs are served by exactly two routes, both with the same costs. In this case both OD pairs have two possible routes that have the same cost.

When it comes to solving the problem of traffic assignment in a dynamic fashion, i.e., considering the time-dependence relationship between traffic flow and travel time, several methods have been proposed. Some are analytical, based on nonlinear programming, optimal control problems or variational inequalities (see Gawron [1998a] for references and Ran and Boyce [1996] for an overview of dynamic models). A frequently used method is based on nonlinear programming, in a similar way that is solved by the Frank-Wolfe algorithm, i.e., subproblems can be solved using a shortest path algorithm for every OD-pair, but adding flow propagation constraints. However, in this case the complexity of the problem increases drastically compared to the static case, because many "copies" of the original network (each can be viewed as a shortest path problem for every OD-pair with an additional flow propagation constraint) have to be handled. Thus, it is clear that these solutions do not scale for large networks. Also, depending on how the dynamics of traffic are described, the dynamics of the flow variables are not consistent with the travel times. For instance, if a function is used to relate volume to travel time (see the concept of VDF in Section 4.4), then the resulting travel times may not correspond to realistic speeds.

## 4.3 CHANGING THE PERSPECTIVE: FROM TRIPS TO DRIVERS

In more recent approaches, explicit treatment of drivers or road users replaces the focus on trips. This is a conceptual or paradigmatic change that was possible by the increase in computational power, and by the progress in microscopic simulation tools. As opposed to the methods where trips are central, methods that model drivers explicitly, allow for individuals to make decisions about their routes or whether or not to change the route. This means that the basis for this approach is that each individual tries to optimize his own, individual travel time. Thus, the key element in the classical DTA/SUE/DUE approach, namely, the ability to determine an arbitrary fraction of trips for which routes will change (see the end of Section 4.2.3), is no longer possible because this is ultimately a decision of the drivers themselves, and not an arbitrary decision of the simulator. Also, as we will discuss in Chapter 8, if these modeled drivers are equipped with devices that enlarge their knowledge about the congestion state of the whole network, then they may be able to perform reasoning or even experimentation that the classical DTA/DUE approach does not necessarily foresee.

In applying the methods of DTA/DUE to active driver assignment, one assumes that each individual has a perfect knowledge of the network state. In the case of commuter traffic, where the traffic demand and hence the network state is more or less the same every day, this assumption is justified. Also, with advanced information provided by in-vehicle navigation system, the context awareness of a driver is increased—possibly to an up-to-date information of the full state of the traffic system. However, then, another issue arises, namely that the value of information may actually decrease if nearly every road user has the same information (see, e.g., Arnott et al. [1991]).

Another assumption made in these approaches is the one about perfect rational and uniform behavior. These assumptions can be relaxed by making a distinction between the actual travel time and the travel time perceived by the individual in previous travels. The perceived travel times are described as random variables distributed across the travelers.

The issue of degradation in performance of decentralized decision making caused by the selfish behavior of individuals remains an important research topic. More recently, Koutsoupias and Papadimitriou [1999] have proposed the so-called price of anarchy (known as PoA) to measure this degradation. The price of anarchy is defined as the ratio between the worst equilibrium and the optimal centralized solution, and is commonly used in data networks. However, contrarily to this kind of networks, which can be monitored and controlled to a given extent, it is impossible to efficiently control vehicular traffic, because drivers and users of the transportation network in general act as "intelligent" and autonomous decision-makers, seeking to maximize their own utility in a bounded rational[3] way, as opposed to data packets that follow some protocol.

Some researchers have created metaphors for explaining the effects of assignment performed in a selfish way, trying to understand the differences of a centralized vs. a decentralized

---

[3]Bounded rationality has been introduced by Simon [1957] as the idea that humans act rationally only within the context of their individual knowledge, experiences and capabilities.

approach. Already in 1968 the Paradox of Braess was introduced, stating that adding a new road to a traffic network may not reduce the total travel time [Braess, 1968].

This "paradox" shows a phenomenon that contradicts common sense: in a traffic network, when a new low-cost link is added to the network (hence an increase in supply), it is possible that there is no reduction in commuting time. Frequently, travel time even increases and so too the costs for the commuters, although one would expect an improvement due to the increased supply, despite no overall increase in the demand. This happens because drivers perceive the low-cost link and rationally try to use it. This creates a new equilibrium that is worse than the previous one. There have been many works that analyze how the Braess paradox works and suggest solutions, for example by allowing drivers to learn the value of their actions [Tumer and Wolpert, 2000].

Game theory also offers scenarios for analyzing human route choice: minority games are scenarios in which agents have to choose between two (or more) alternatives. The agents that select the option with the least number of agents are better off. Another work based on minority games is the one by Galib and Moser [2011], who have investigated the use of this approach to achieve a balanced usage of a road network in which decision are made about which link to follow. These scenarios clearly mirror the situation in traffic networks when the route with more users is congested, leading to increased travel time for those who choose it. Extensive research has been conducted using two-route scenarios with different particular setups for analyzing how rational or bounded rational agents can autonomously find equilibrium distributions [Chmura and Pitz, 2007, Klügl and Bazzan, 2004b]. These scenarios have also been extensively used for analyzing the impact of traffic information, depending on how many agents follow the advice Klügl and Bazzan [2004a], even with information that is manipulated to drive the agent's decision making towards the system optimum.

## 4.4   EVALUATION OF ROUTE ASSIGNMENT

In the iterative cycle in Figure 4.1 a central element is the evaluation of the assignment. In the simplest case, a link performance function is used, which directly relates load to speed and thus to travel time and delay. Such functions are derived from fundamental diagrams introduced in Chapter 2. Typically, the travel time on each link can be described by a so-called volume-delay function (VDF), a function that expresses the average or steady-state travel time on a link depending on the volume of traffic on this link. Different traffic assignment tools use different VDFs. Also, different VDFs exist for different kind of traffic: traffic on freeways, non-urban roads or within urban systems. A well-known VDF for freeway traffic is the one suggested by the BPR (Bureau of Public Roads), as given in Eq. 4.2. There, $t_0$ is the travel time at free flow, $V$ is the volume (in vehicles per hour), and $C$ is the capacity. $\alpha$ and $\beta$ are factors that have to be empirically determined for adapting the formula to a particular kind of road segment determining how the function reacts with increasing volume. This VDF consists of a component for travel time with free flow speed and a component describing the effect of congestion.

$$t = t_0 \left[ 1 + \alpha \left( \frac{V}{C} \right)^{\beta} \right] \tag{4.2}$$

The VDF in Eq. 4.2 is one of the simplest formulas; corresponding VDFs for country roads need to consider the chances for overtaking in single-lane roads. For urban traffic, the free flow travel time is only secondary; the travel time on a link is rather governed by waiting times at crossroads. Cascetta [2009] gives a short introduction, while other VDFs can be found in Chapter 10 of Ortúzar and Willumsen [2001].

VDFs simplify the calculation and thus the task of finding the network equilibrium. Perhaps due to these characteristics, VDFs are widely used to characterize link performance in works such as the aforementioned Braess paradox [Braess, 1968], and those derived from the idea of the price of anarchy as in Koutsoupias and Papadimitriou [1999] and Roughgarden and Tardos [2002]. VDFs can be only used when the dynamics of the assignment are disregarded as they abstract away all dynamics on the road. Thus, it is not difficult to see that evaluating assignment using VDFs has limitations. First, VDFs allow links to have a volume/capacity ratio greater than one. Consequently, a second limitation is that VDFs do not allow the representation of the phenomenon of congestion spillback. Although this is implicitly represented by the possibility of volume being higher than capacity, this inflated volume refers only to the particular link, so no physical spillback is actually represented. This means that congestion on the current link does not affect upstream links. Third, VDFs assume a kind of first-in-first-out scheme and therefore do not account for overtaking. Consequently each vehicle on the link experiences the same travel time. Fourth, it is hard to account for different kinds of lanes: lane-based effects such as high-occupancy vehicle lanes or high occupancy toll lanes cannot be accommodated.

To address these problems of using VDFs, another family of solutions for evaluating the effects of a particular traffic assignment is based on simulation, using approaches discussed in Chapter 6. The basic idea is to select some initial set of routes for each OD pair, assuming travel time under, for example, free flow situations. After routes are assigned for each OD pair, the network load is simulated by actually reproducing the traffic flow produced by the trips of all road users. Based on this, travel times are determined. Although such simulation-based assignment evaluations can be done in a computationally efficient way, they have the drawback that it is not easy to make statements about convergence and stability of the results. Additionally, further assumptions about road users behavior, such as departure time, individual desired speed, etc., have to be included. Clearly, the level of complexity involved relates to what type of simulation paradigm is used (see Chapter 6).

## 4.5   TO KNOW MORE

The stochastic user equilibrium is a classical method in traffic assignment. Readers interested in variants of the basic model [Daganzo and Sheffi, 1977] and stability issues are referred to Hazelton [1998] and Watling [1999]. Comprehensive surveys on traffic assignment can be found in [Sheffi, 1985] or [Patriksson, 1994] for static assignment and Peeta and Ziliaskopoulos [2001] or [Chiu et al., 2011] for dynamic assignment. Also, [Ortúzar and Willumsen, 2001] offers a comprehensive introduction to traffic assignment in Chapters 10 and 11.

Research on the value and role of traffic information provision turns more and more important. Apart from the already quoted work by Arnott et al. [1991], readers may also be interested in taking a look at Levinson [2003] and dePalma et al. [2012].

CHAPTER 5

# Getting Data for Demand Estimation and Traffic Flow Modeling

The main aims of data collection in traffic engineering are to gather data to forecast future transportation needs for planning purposes; to forecast impacts of measures taken to improve efficiency; and to improve the understanding about parameters of the traffic flow model (as for instance how density changes with speed and flow). Section 5.2 addresses the latter. The other two relate more generally to demand estimation and are discussed next.

## 5.1 DATA COLLECTION FOR ESTIMATION OF DEMAND AND VOLUME

Data required to estimate demand and volume refer basically to countings (of vehicle, passengers, etc.). Countings can be obtained by various devices/methods. From these countings, some quantities introduced in previous chapters (volume, flow, demand, and capacity) are derived. Although all of these are expressed in the same unit (for example number of vehicles per time period), they are not the same. We also remark that frequently, the words volume, flow, and even demand are used interchangeably.

Demand differs from volume in the sense that the former is the number of vehicles (or passengers) that *desire* to travel. In some cases, demand can be higher than the volume. Capacity is a characteristic of the facility itself; it is estimated based on procedures such as those described in the Highway Capacity Manual [Tra, 2001].

While volume and flow can be counted, demand is more difficult to estimate, as it requires knowledge about the number of road users who have changed their routes to avoid a (congested) location in which the demand is higher than the capacity, as well as the number of road users who have dropped their trips altogether. Demand in arterials is more difficult to estimate that demand in freeways, due to the various diversion points (intersections) that constitute alternative routes.

To further complicate the problem, a characteristic pattern of traffic volume is its concentration on certain hours of the day and certain days of the week. Strong peaks may occur, e.g., when commuters are traveling to and from their workplaces. According to Roess et al. [2004], depending upon region and location, the peak hour typically contains 10–15% of the 24-h volume.

Given this characteristic of seasonality and within-day variation, a question remains: what should be measured. Volumes can be measured for different time periods (e.g., 1–5 min) and then extrapolated. However, it seems that in order to be statistically significant, the shortest period of time over which volumes should be measured is 15 min [Roess et al., 2004]. Further, these authors make the point about one study that has measured a flow of 2200 veh/h over 5 min, 2050 veh/h over 15 min, and a peak hour volume of 1630 veh/h. The second value is 20% higher than the peak hour, which could have a huge impact if used, for instance for purpose of designing signal timings. Similarly, there are differences in daily patterns, as well as in patterns of road used for different purposes.

At this point we would like to remark that concentration of traffic in well-known patterns is less and less a typical phenomenon in modern societies. A "typical" morning peak in moderns mega-cities looks more or less like this: after 10 am, the traffic remains relatively high, showing that afternoon peaks no longer exist. This is a phenomenon caused by (a) individuals making travel choices that allow them to travel "off-peak" hours and (b) changes in the business nature and locations that are more spread. These processes may cause peaks and off-peaks periods to be virtually indistinguishable. Despite this fact, no matter the pattern, volumes need to be measured.

To count the volume of vehicles or passengers that pass a given point, various techniques and devices can be used, which vary greatly in their complexity and cost. The simplest way is to use manual counts. Besides, this method can be used when automated procedures have difficulties discerning movements and/or classes as for instance turning movements at intersections or vehicle classification. Another category of count devices are those that are portable such as pneumatic road tubes.

Permanent count devices (which are normally coupled with a data communication device and a microprocessor) normally transcend the task of just counting volumes. Rather, they are employed for real-time monitoring and control (e.g., of traffic signals). These devices are mostly magnetic-loop detectors, but there are other techniques such as cameras and sonic and ultrasonic detectors. In any case, volumes counted can be presented as in Figure 5.1, which depicts manual countings for an intersection of two arterials in the city of Porto Alegre (Brazil), collected (over 15 min and extrapolated to a period of 1 h) on June 6, 2013 from 08:26 to 08:41 am.

Due to their wide use today, we focus on inductive loop detectors. This kind of detector is installed in the road surface (e.g., by saw cutting in various shapes). The loop is laid into the cut, sealed, and connected to an electrical source. This creates an electromagnetic field that can detect metallic objects such as vehicles moving across the loop. A single-loop detector can directly measure the time at which the front and the rear of a vehicle passes the detector. Unless one assumes an average vehicle length, it is not possible to measure speed of a vehicle using a single-loop detector; in order to be able to do so, a double-loop detector is necessary.

More sophisticated loop detectors may aggregate data from each single vehicle data by averaging over fixed aggregation time intervals (typically 1 min). This aggregated data is then

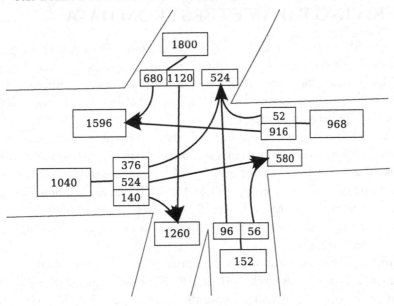

**Figure 5.1:** Example of vehicle countings for an intersection in Porto Alegre (AM peak).

transmitted to some control facility. This way, it is possible to measure the flow passing a cross section of the road over a time interval.

Another measure that is frequently used is the occupancy of a detector: the fraction of the aggregation interval during which the cross section of the road is occupied by a vehicle.

Normally, studies related to demand require data related to OD matrices, as seen in Chapter 3. Such data generally involves surveys and other kinds of interviews in order to establish some travel patterns. Interviews could be made at home, by phone, or by stopping the vehicle and asking the travelers a short series of questions regarding the origin, destination, purpose, route, and eventually, modes involved in that trip. Alternatively, interviews could be carried out at centers that knowingly attract trips (shopping centers, etc.) in order to reduce the scope of the questionnaire.

Apart from the surveys, some ITS technologies mentioned in Section 1.3 may help to provide data for the determination of an OD matrix. For example, automated license-plate reading (originally conceived for speed and traffic law enforcement), automated toll collection (in areas with a dense network of toll collection points), RFID, and even in-vehicle navigation devices and mobile phones could be used for such data collection. More on these methods appear in Chapter 8. However, one important issue with these methods based on navigation devices is privacy.

## 5.2  DERIVING PARAMETERS FROM DATA

Observation of traffic flow is an important step in the task of estimating the parameters of any traffic flow model. To do this, there are some methods and ways reported in the literature (instantaneous, local, quasi-local, etc.). For a deeper discussion on the differences and effects, see Leutzbach [1988]. An introduction to this topic can be found in Kerner [2009], Leutzbach [1988], as well as Treiber and Kesting [2013].

The most traditional observation method relates to keeping continuous records of streams at a fixed measuring point (normally a cross section of a road or lane) during a given time interval. If one records the accumulated count of vehicles $K(t)$ at this measuring point as a function of time $t$, then the volume of traffic is given by $K(t + \Delta_t) - K(t)$. However, flow is defined as number of vehicles over a certain time interval. Thus flow is given as $q = \frac{K(t+\Delta_t)-K(t)}{\Delta_t}$. Normally these countings are non-stationary, i.e., time-dependent, thus posing the challenge that the quantity $\Delta_t$ must be selected carefully.

It is difficult to get a continuous measurement of instantaneous speeds of vehicles moving along the road, i.e., a continuous tracking of each vehicle speed. Therefore, measurements of speed are made at a given point of the road, only for those vehicles passing this point. This is called the local measurement of speed. This means that the speed of each vehicle is measured only once, exactly when the vehicle passes this point. The average speed (of several vehicles, over a time frame $\Delta_t$) is then computed as an approximation for the instantaneous speed of a given vehicle at that point.

While flow, occupancy (of a detector located in a cross section), and the average speed can be measured directly, other macroscopic parameters discussed in Section 2.2 can only be estimated by making some assumptions. The case of density is the most prominent. It is defined as the number of vehicles on a given road segment but detectors located in cross sections can only measure temporal averages.

Also, data that is inherently related to microscopic parameters can be collected. For instance, cameras installed in tall buildings can track single vehicles and it is possible to determine their trajectories. However, according to Treiber and Kesting [2013], camera-based methods involve complex and error-prone procedures that require automated and robust algorithms for the vehicle tracking, and thus are often the most expensive option for data collection. Furthermore, a camera can cover a road section of at most a few hundred meters, and it is difficult to observe smaller vehicles hidden behind larger ones.

Currently, without the use of sophisticated methods, it is possible to capture trajectory data of a relatively small amount of vehicles within a given cross section of a road, in a given time period. This allows the determination of microscopic parameters and, from them, data about traffic density.

Trajectory data is normally presented in spatio-temporal diagrams such as the one in Figure 5.2. Each line in this diagram is a trajectory of a single vehicle. To exemplify how this works, we use a scenario that was simulated. This means that we simulate the task of gathering the data by means of detectors. These trajectories in particular refer to vehicles traveling in the circular road discussed in Section 2.3. We show the trajectories of 90 vehicles traveling in that circular road. For each vehicle, we have recorded its position and plotted it against travel time (we use a single clock, thus time is relative to this clock). Besides the trajectories themselves, these diagrams allow the visualization of the speed at a given position and time (given by the gradient of the trajectory) of each vehicle. In particular, trajectories with no variation in the vertical (distance) axis correspond to standing vehicles. Also, it is possible to determine the time headway between two vehicles (horizontal distance between two trajectories), and the distance headway between two vehicles (vertical distance between two trajectories).

Furthermore, macroscopic parameters can be estimated, such as flow (number of vehicles passing a given location during a certain time) by counting the number of trajectories crossing a horizontal line within a time interval; and density (number of vehicles counted in a given road segment at a given time), by counting the number of trajectories crossing a vertical line. Finally, it is possible to visualize the propagation of a traffic jam. Congestions in Figure 5.2 are stop-and-go waves that are moving upstream. These waves can be better observed by taking a bird's-eye view of the roadway. A spatio-temporal diagram is then a way to reproduce this view of the wave. These waves are also known as shockwaves because they are transition zones between two traffic states (flowing and congested), which move through a propagating wave.

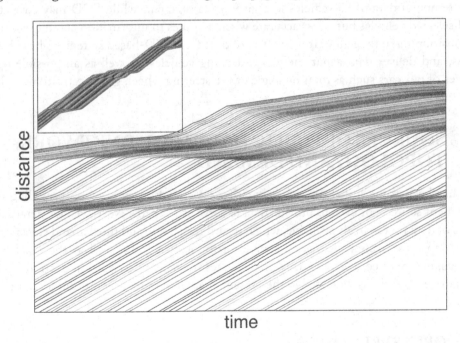

**Figure 5.2:** Example of spatio-temporal trajectories in which some shockwaves "move" (the main plot is a zoom of the inset plot).

There might be other purposes for recording vehicles's trajectories; for instance, for estimating travel time and monitoring congestions. A promising approach for this is the use of GPS traces, with the advantage that (i) there is no need to process images and (ii) it applies to massive amounts of data. While there are some issues related to this technique as, e.g., privacy and map-matching the geo-referenced coordinates, the literature also reports several initiatives centered on so-called probe vehicles that "float" in the traffic flow. These are special vehicles such as taxis or trucks dedicated to delivery and other commercial purposes. This type of data is called floating-car data (FCD). So far this type of FCD is of limited use since (i) they do not cover the whole network and (ii) are not fully representative due to their relatively low speeds. Nevertheless, if the purpose is to measure speed of main arterials under congested situations, then FCD data is very useful and is indeed being used. In Stuttgart, for instance, a fleet of approximately 700 taxis (data from 2003) circulate with GPS device that send its position and speed to a control center approximately every 30 s. This center collects information from this and other sources in order to feed a monitoring and control system. Similar projects are operational in cities such as Vienna, Berlin, and Nürnberg. Future versions of FCD (called extended FCD or xFCD by some researchers) must include some way to anonymize the trajectories collected from standard navigation systems that are turning more and more popular.

In summary, trajectory data and FCD differ in the sense that the former records detailed spatio-temporal data of all vehicles in a given road segment, while FCD may cover the whole trip of selected vehicles but is less accurate when it comes to information about exact trajectories (e.g., positioning in the lanes). It is a matter of time for xFCD-based systems to be able to record, process, and deliver data about the gap to leading vehicles, as well as all possible operational variables of our cars such as rotation angle of the steering wheel, gears, activation of signals and lights, etc.

## 5.3 SOURCES OF PUBLIC DATA ABOUT NETWORK TOPOLOGY AND DEMAND

There has been some initiatives that rely either on crowdsourcing, or on providing data from government sources, or even to provide data to serve as testbeds. Next, we give details about two sources of data that is freely available, which regard the supply and the demand sides: while Open Street Map (next section) is a good source for topological data (such as the exact location of junctions, nature of connecting edges, etc.), the test instances described in Section 5.3.2 provide OD matrices, which, as discussed in the present chapter, require a significant amount of work to generate.

### 5.3.1 OPENSTREETMAP

According to http://wiki.osmfoundation.org, OpenStreetMap (OSM) is an initiative that has started in 2004 to create and provide free geographic data (such as street maps), to anyone.

The international foundation (registered in England) that supports the OSM project is dedicated to encouraging the growth, development and distribution of free geospatial data, and to provide geospatial data for anyone to use and share. OSM was inspired by the success of Wikipedia, crowdsourcing, and the preponderance of proprietary map data. Around one-third of the over 1 million registered users (as early 2013) have contributed at least one point to the OSM database. Users can collect data using GPS devices, aerial photography, data from governments, and other sources. These data have been favorably compared with proprietary data sources (see Zielstra and Hochmair [2012] and Sehra et al. [2013]) though data quality varies worldwide [Haklay, 2010].

Map data is usually collected using a GPS unit and is uploaded into the database. Afterwards, it has to be annotated using a map editor (e.g., the tool JOSM) to add information about the kind of path that refers to given GPS data as, e.g., a motorway, footpath, or river. Other information such as placing and editing objects such as schools, hospitals, shopping and other businesses, bus stops, etc. is done based on contributors local knowledge.

OSM's' license was changed recently (September 2012) to the Open Database License (ODbL). More information about license, commercial users, data format, and software for editing maps can be found at `http://en.wikipedia.org/wiki/OpenStreetMap` and `http://wiki.openstreetmap.org/wiki/Beginners_Guide_1.1`.

It is possible to download all, or parts, of the database, but mostly the use is via accessing the OSM website at `osm.org` and selecting a rectangular portion of a map. This action generates an XML formatted .osm file. This kind of format/file is currently recognized in a series of free software initiatives such as the traffic simulator SUMO (see Section 6.1.2).

### 5.3.2 TRANSPORTATION NETWORK TEST PROBLEMS

The website `http://www.bgu.ac.il/~bargera/tntp/` at Ben Gurion University collects transportation test problems that are available for anyone who wants to test routing and assignment algorithms. It provides data about maps and demands for some cities in the world. This data refers to network topologies and demands (trips) to be assigned. Normally, this data is divided into two or more kinds of files, e.g., one to describe the topology and another containing the trips.

For example, the data about the city of Sioux Falls is provided in three files. One contains the network (SiouxFalls_net), a second one contains the coordinates X and Y of the nodes of the network (SiouxFalls_node), and a third one describes the OD (SiouxFalls_trips). In some cases, the former file brings also information about costs (e.g., toll), free flow travel time, and values for parameters associated with VDFs (see Section 4.4).

## 5.4 TO KNOW MORE

Following a general trend, data fusion is becoming a hot topic in the field of transportation. Indeed, combining data from multiple, heterogeneous data sources such as cross-sectional data, floating-car data, "floating-phone data," police reports, etc., is mentioned in Treiber and Kesting [2013] as a challenging task. This challenge arises because each of these categories of data describes different aspects of the traffic situation and might even contradict each other. However, data fusion in fact goes beyond these issues. In Chapter 9 we discuss how this is being extended to include data from social networks in order to estimate the traffic state. Also, in see Section 8.4 for a brief discussion on how GPS-based tools can be used instead of traditional surveys, in order to have access to more detailed travel diaries that are useful in activity-based demand modeling.

# CHAPTER 6

# Modeling and Simulation of Advanced Decision Making

Modeling and simulation is one of the major application areas for computer science in the transportation domain.

A large variety of existing traffic simulators have demonstrated their value in the practice of traffic management and operations. Reproducing traffic systems in a model is an established means of predicting the effects of a particular political or infrastructural measure. Traffic simulators are not only used to test the impact of new infrastructure—such as a new bypass road or the suitability of a particular signal plan. Simulation is also used to test the feasibility of technical innovations such as new communication or localization systems. In (immersible) driving simulators, traffic simulation is used to generate the traffic context for the driving task.

It is beyond the scope of this text to survey and compare traffic simulators, whether commercial or not. The interested reader is referred to a collection of chapters by Barceló [2010]. Also, Passos et al. [2011] discuss different simulators with respect to their suitability for simulating urban traffic systems of the future.

Traffic models have existed for decades. Traditionally, their origins can be found in physics, mathematics, and statistics. During the last 20 years, different techniques from AI and computer science in general have found their way into traffic simulation. Since the middle of the 1990s, agent-based simulation has become a relevant umbrella for different approaches to traffic simulation. Agent-based models use the concept of multiagent systems as a underlying metaphor for the model [Klügl and Bazzan, 2012]. For traffic simulation, this means that the decision making entities (e.g., travelers or households) actively decide themselves about their activities or interactions, based on their local information. This is a paradigm change, as discussed in Section 4.3, which is now possible given the advances in computational power. In practice, this means that it is now possible to simulate each individual deciding locally about his departure time, about mode of transportation, about his route, etc. In these decision making processes, models from cognitive science can be used.

Many different approaches to traffic simulation exist, with different levels of granularity reproducing the intelligent behavior of drivers to a greater or lesser extent. In Chapter 3 we tackled modeling the intelligent decision making with respect to daily activities, containing intentional selection of destinations (mainly secondary destinations for shopping and leisure activities). Also, the abstract link performance functions (VDFs) introduced in Chapter 4 constitutes a model that aggregates the actual traffic flow over a given interval of time. In the following, we shortly characterize standard approaches and then continue with AI and agent-based models for traffic simulation.

# 6.1   SYSTEMATICS OF TRADITIONAL APPROACHES

Table 6.1 (from Treiber and Kesting [2013]) gives a methodology for model-based analysis of traffic systems. A main distinction must be made between simulation for traffic planning and traffic flow simulation (see Treiber and Kesting [2013]). Three orthogonal dimensions, namely (i) temporal granularity, (ii) area of application, and (iii) aspects that are related to the core questions addressed when modeling, are used as follows.

- The temporal aspect: the timescale in traffic flow dynamics is of the order of minutes to a few hours, while transportation planning covers periods from hours to several days or even years.

- While traffic flow dynamics assume an *externally* given traffic demand and fixed infrastructure, planning considers changes in the infrastructure and its effects.

- Traffic flow dynamics analyzes human (or automated) operational driving behavior (accelerating, braking, lane-changing, turning), while higher-level actions—e.g., activity choice (number and types of trips), destination choice, mode choice, and route choice—belong to the realm of transportation planning.

**Table 6.1:** Categories of traffic modeling and simulation [Treiber and Kesting, 2013]

| Time Scale | Area | Model | Relevant Aspects |
|---|---|---|---|
| < 0.1 s | | Sub-microscopic | Motor control, brake system |
| 1 s | | Car-Following Models | Reaction time |
| 10 s | Traffic Dynamics | Car-Following Models | Accelerate, decelerate |
| 1 min | | Car-Following Models | Traffic signal cycle time |
| 10 min | | Macro Models | Stop and go waves |
| 1 h | | Assignment | Peak hour |
| 1 day | Traffic Planning | Traffic Demand | Daily traffic behavior |
| 1 year | | Traffic Demand | Infrastructure measures |
| 5 years | | Statistics | Spatial planning |
| 50 years | | Prognosis | Demographic change |

While techniques and approaches relevant for traffic planning have been already discussed in Chapters 3 and 4, traffic flow simulation has only been shortly tackled with respect to its application for predicting travel times (evaluating the assignment of vehicles to routes as in Sections 2.3

and 4.3). As mentioned in those sections, instead of directly connecting load and travel time with help of a VDF, the traffic flow can be simulated, so that the travel time is rendered from the actual interactions.

Depending on aspects of aggregation level, mathematical structure, and conceptual aspects, different models of traffic flow can be evaluated. In Table 6.1 microscopic and macroscopic models appear as the main categories. To these, also sub-microscopic and mesoscopic models can be added, as discussed in the next sections.

## 6.1.1    SUB-MICROSCOPIC MODELS

Sub-microscopic models deal with the actual functioning of a vehicle as modeled by the laws of physics. They allow for analyzing the interaction between the different components within the vehicle in reaction to the actions of a driver when performing actions such as braking, accelerating, or turning the steering wheel. First principles, physical processes are the main ingredients for the model.

## 6.1.2    MICROSCOPIC TRAFFIC FLOW MODELS

Microscopic models describe the behavior of an individual "driver-vehicle unit" (DVU) as a reaction (accelerating, braking, lane-changing) to the surrounding units. We remark that in the literature stemming from the physics area, DVUs are also called "driver-vehicle particles." The simulation of these units and their interactions produces the overall traffic flow. Variables of the model are the individual units' positions, speeds, and acceleration rates. Microscopic models are useful for the following objectives:

- Modeling how a single DVU affects the traffic system, with individual decisions based on information from ATIS (Section 1.3), from advanced driver-assistance systems, as well as from information resulting from using innovative communication systems for interacting with infrastructure or other vehicles (see Chapter 8).

- Testing the efficiency of particular measures of traffic optimization on heterogeneous traffic participants, e.g., when simulating the effects of speed limits or bans on passing for trucks.

- Modeling the particularities of human driving behavior, including estimation errors, reaction times, inattentiveness, and anticipation.

- Visualization of interaction between various traffic participants (cars, trucks, buses, cyclists, pedestrians, etc.).

- Producing background traffic for driving simulators in different applications such as research on traffic psychology and behavior, or serious game simulators for training purposes.

Microscopic models may take many different forms, ranging from simple rules to models mirroring drivers' behavior. In principle, two elements of drivers behavior have to be addressed in a microscopic traffic simulation: (i) driving with adaptation of individual speed and (ii) lane changing behavior and overtaking. Eventually, a third element is added: route choice behavior. In the following we concentrate on the former element (actual driving). Route choice from a driver perspective was covered in Chapter 4. Here, we remark that when the model does not explicitly consider route choice by the DVUs, turning probabilities are assigned to the intersections. This allows for the simulation of the DVUs traveling through the network.

An important class of microscopic traffic flow models is the one composed by **car-following models** Gipps [1981], which aim at generating the simulated traffic flow based on local interactions. In the simplest case, the speed of a vehicle is modeled as a function of the distance to the vehicle in front, assuming fixed values for acceleration or deceleration. More sophisticated and realistic models regulate acceleration instead of speed and are not only based on distance, but also on the speed of the preceding vehicle. The so-called "Intelligent Driver Model" [Treiber et al., 2000] (shown in Eq. 6.1, following Treiber and Kesting [2013]) is a well-known example in traffic engineering, although from an AI point of view its level of intelligence is restricted. In this equation, factor $a$ denotes the acceleration; $v_0$ is the desired speed; $\delta$ is an exponent; $v$ denotes the current speed; $s$ is the current distance to a leading DVU; and $s^*$ is the acceptable gap between the DVU and the DVU ahead, which depends on the former's current speed and on the difference between their speeds. Parameters of the behavior description ($a$, $v_0$, and $\delta$) can be used to characterize the individual driving style as well as given speed limits [Kesting et al., 2010].

$$\dot{v} = a \left[ 1 - \left( \frac{v}{v_0} \right)^{\delta} - \left( \frac{s^*(v, \Delta v)}{s} \right)^2 \right] \tag{6.1}$$

Barceló [2010] contains a collection of other simulations models that are used in established commercial and non-commercial microscopic traffic simulation systems. As of 2013, an important representative of a commercial simulator is VISSIM (see www.vissim.de, [Fellendorf and Vortisch, 2010]). It was further developed into a multi modal traffic flow simulator that includes public and freight transportation as well as pedestrian movements in addition to vehicular traffic. Other examples of commercial simulators using microscopic models are PARAMICS (http://www.paramics-online.com/) and AIMSUM (http://www.aimsun.com). An open source simulator is SUMO ("Simulation of Urban Mobility," http://sumo-sim.org/, [Krajzewicz et al., 2012]), which has several microscopic models built in for testing.

The aforementioned car-following models are continuous, i.e., they represent driver behavior as if fully corresponding to an automatic cruise control systems. Microscopic traffic flow models that are fully discrete, in time, space and state, use cellular automata. For this purpose, cellular automata are based on similar concepts of car-following, yet are more abstract as there are no explicit moving entities. Rather, the state of a cell represents whether there is a DVU in this road segment, and, in affirmative case, which is its speed. what speed this "state-vehicle" travels with. In principle, a cellular automata based model is an intermediate type of model, being originally based on discretization of partial differential equations of fluid flows. Cellular automata traffic models can be implemented efficiently and are also easy to parallelize.

A well-known, seminal cellular automata model is the Nagel-Schreckenberg model, originally introduced by Nagel and Schreckenberg [1992] for freeway traffic. The road is segmented into cells of a size resembling the effective length of a vehicle. The state of each cell is updated depending on the state of the neighboring cells. The behavior of a cell in the Nagel-Schreckenberg model is formulated by means of the following rules:

1. Deterministic acceleration depending on the number of empty cells ($g_\alpha$) in front of a DVU $\alpha$, as well as its maximum speed: $v_\alpha(t + 1) = \min(v_\alpha(t) + 1, v_0, g_\alpha)$ with $v_\alpha(t)$ as the speed in cells at time $t$, and $v_0$ as the desired speed. In Nagel and Schreckenberg [1992] this is formulated as two rules, one for acceleration for reaching the desired speed and one for potential deceleration depending on the length of the gap to a leading vehicle.

2. Random decrease of speed: With probability $p$, $v_\alpha(t + 1) = \max(v_\alpha(t) - 1, 0)$.

3. The state change resulting from the actual movement.

In this model, under free flow, acceleration is set to 1 per time-step increasing speed up to a maximum value. Deceleration is done to avoid a crash when the regime is not free flow. In order to reproduce real-world traffic flow (where for instance unmotivated distraction and other phenomena happen), rule 2 simulates random deceleration. This stochastic behavior element is responsible for reproducing traffic jams out of nothing, which are frequently observed.

Originally formulated as a 1-dimensional ring, the model has several extensions, e.g., for lane changes. It was mainly applied to freeway traffic for online simulations (see Wahle et al. [2001]). Although the quite coarse temporal (usually $\Delta t = 1$ second) and spatial resolution makes the model less appropriate for urban networks with many intersections and short links, it has been adapted for use in urban scenarios (see, e.g., Esser and Schreckenberg [1997]).

A more elaborate cellular automata model is the Kerner-Klenov-Wolf Model [Kerner et al., 2002], in which a speed synchronization is explicitly integrated. Within a certain distance interval to the preceding vehicle, a vehicle adapts its speed, resulting in some form of steady state in which there is no direct relationship between distance between vehicles. Kerner et al. [2002] show that their model is more realistic concerning different empirically observed phenomena in traffic flow.

Both continuous car following models and discrete cellular automata models are based on the representation of local interactions. The basic entity—a DVU or a cell—can be seen as a "particle" that adapts its speed to the speed of the surrounding "particles," based more on fluid dynamics and statistical physics than on psychology focusing on human behavior. Hence, they resemble traffic as a system of locally interacting particles rather than as a simulation of individual human driving behavior with explicit reaction time, perception failure, problems in estimating distances or speed, problems with attention and distraction. Additionally, only the directly preceding vehicle is relevant for adapting the vehicle's speed. There is no deliberation or anticipation of the preceding car's behavior or vehicles further ahead. After finishing this section with other traditional traffic simulations, we will tackle new approaches in the next section that explicitly deal with the reasoning of drivers (rather than just particles as homogeneous DVUs), with the intention to create models with more human-like characteristics.

### 6.1.3   MACROSCOPIC TRAFFIC FLOW SIMULATION MODELS

Macroscopic models describe traffic flow analogously to liquid or gases in motion. As mentioned in Section 2.2, the dynamical variables are locally aggregated quantities such as the traffic density, flow, mean speed, or the speed variance. Because the aggregation is local, these quantities may vary not only in time, but also across different spatial units. Due to this abstract and aggregated view, macroscopic models are faster to calculate and run. Also, the calibration task is restricted to less parameters. Thus, if the effect of model elements such as lane changes, heterogeneous driver-vehicle types (that are difficult to describe macroscopically) are not relevant for the simulation objective, or, one is dealing with analysis of a large network, then it is advisable to use only macroscopic quantities, i.e., a macroscopic model should be used [Treiber and Kesting, 2013].

The basic principle behind macroscopic models is the conservation law for vehicles, shown in Eq. 6.2 [Treiber and Kesting, 2013]. It represents the idea that temporal changes in the density of vehicles over time may only originate from changes in the flow coming from neighboring sectors or going to neighboring sectors:

$$\frac{\partial k}{\partial t} + \frac{\partial (kv)}{\partial x} = 0 \ . \tag{6.2}$$

The scenario-specific spatial geometry and the particular mathematical formulation of the conservation law determine the particular form of a macroscopic model. In general, the whole road segment is divided into smaller sectors of an appropriate size. For a reasonable aggregation, each sector must be large enough so that a sufficient number of vehicles fit in it. However, it must be small enough so that the assumption of a homogeneous value (e.g., speed) remains valid. The main differences between the different macroscopic models come then from the way in which speed or flow are modeled.

One of the simplest models is the Lighthill-Whitman-Richards (LWR) model; it has only one dynamic equation, which computes the density of vehicles. Treiber and Kesting [2013] give an introduction into the Cell-Transmission model, which is a discretization of the LWR model in time and space. It is based on a simplified fundamental diagram (see Section 2.3). Parameters of the model, such as desired speed or maximum density needed for determining the fundamental diagram are taken from statistics for different types of roads.

In contrast to this, in METANET [Kotsialos et al., 2002] both density and mean speed are dynamic and calculated for each segment of a link in a road network. The conservation law for vehicles just mentioned updates the density. The mean speed update for segment $i$ contains a density-dependent component for segment $i - 1$ (upstream), a component that adapts to speed differences of inflowing vehicles, and a component that reacts to density effects on speed in the downstream segment. METANET has not just been applied for simulating network traffic, but it was used especially for predictive control. In this type of control, a model is used for predicting the effects of a control adaptation in an online way. Consequently, the model must combine sufficient accuracy with fast execution.

Macroscopic models also exist for urban networks, although homogeneity conditions for segmentation of links are more difficult to achieve. Also, interactions between different links need to be considered, as for instance queues in one link blocking upstream links and intersections. The assumption of vertical queues—which means that queues have no explicit relation to the length of the link, as if all vehicle were stacked vertically—is also not acceptable for an urban system. The freeway METANET model for urban traffic was extended by van den Berg et al. [2003], with the goal of providing a model for use in predictive control for integrated freeway and urban networks. The part that refers to the urban network explicitly takes that interaction between links into account. The state of a link represents the length of the queue (number of vehicles) on a link. At each intersection, flow is just treated as if none of the connected links were full (blocked).

## 6.1.4 MESOSCOPIC TRAFFIC FLOW SIMULATION

Mesoscopic models are not explicitly listed in Table 6.1, yet they fill a gap between macroscopic and microscopic models combining advantages from both, such as efficient calculation with a finer level of detail and expressiveness. Often, they are combined with a simulation-based procedure for dynamic traffic assignment, because a mesoscopic model can be simulated fast, even when route choices of the vehicles is explicitly considered (for example, see Gawron [1998a]).

There are different types of mesoscopic models:

- **Hybrid models** explicitly combine approaches from both macroscopic and microscopic into a new model. Areas of interest are modeled on a microscopic level of detail, whereas less important parts are modeled macroscopically. These then determine the input to the more detailed one. Examples can be found in Sewall et al. [2011]. Yet, these combined models often suffer from problems related to consistency in the level of traffic dynamics, or to compatibility of route choices, especially at the interface areas between the microscopic and the meso/macroscopic model [Burghout, 2004].

- **Genuine mesoscopic models** use an intermediate abstraction level between the aggregated macroscopic and microscopic levels. For example, individual vehicles are modeled explicitly as entities in the simulation, but instead of determining their current speed based on the distance to the preceding vehicle, mesoscopic models use a density-dependent speed value determined using macroscopic formulas, which in turn are based on the number of vehicles in a road segment. Such principles are applied for example in CONTRAM [Taylor, 2003] or DynaMIT [Ben-Akiva et al., 2010]. Other models are based on the paradigm of queuing systems. Queues of vehicles build up on links, whereas nodes are modeled as servers that transport the vehicles to the corresponding next link. Models such as MEZZO [Burghout et al., 2005], DYNAMEQ [Mahut and Florian, 2010], or MATSim [Balmer et al., 2009] are event-based and differ in how the speed (and thus, travel time) on the link are determined. This difference may arise from issues such as whether there is a distinction between the part of the link on which vehicles move and the part on which the queue is located, and whether the transportation through the nodes happens in a deterministic or stochastic way.

All these traditional models—submicroscopic, microscopic, macroscopic, and mesoscopic—are based on a rather high abstraction of the driver behavior. They are established for their particular application areas. High effort has been invested to validate them based on empirical measurements (see Chapter 5). Hoogendoorn and Bovy [2001] survey traffic simulation models also with respect to their usability.

## 6.2   HUMAN-LIKE DRIVING WITH ADVANCED DRIVER MODELS

The application of AI in traffic modeling and simulation promises to generate more realistic and detailed models of human decision making in traffic, allowing a more precise and reliable prediction of traffic states. These models contain entities with much more detailed reasoning methods than the above described microscopic traffic simulators, thus AI-based models will hardly be as fast in calculation as the simpler ones, based on, for instance, fluid dynamics.

These models are often called "agent-based traffic simulation." Clearly, they belong to the category of microscopic traffic simulations as every decision making entity is modeled explicitly. Kesting et al. [2010] even go so far as to characterize all microscopic traffic flow models as agent-based ones. The interested reader can find further references on agent-based traffic simulation in Bazzan and Klügl [2013].

Agent-based traffic simulation models can be found in all areas of decision making relevant to traffic, ranging from decisions about speed, lane or route choice to selection of departure time or destination choice. For example, in Chapters 3 and 4, we discussed how the adoption of an agent perspective was fruitful for activity-based demand modeling, or for agent-based routing models. In the following, we shortly discuss models aimed at more realistic simulation of human participants in traffic. In general, one can observe a recurring architectural pattern: a layered architecture as shown in Figure 6.1.

The basic idea behind this pattern is that there is a separation between the actual driving, which is modeled using a traditional microscopic or mesoscopic traffic flow simulator and a high-level reasoning layer. The former determines how the driver agent acts on the road, often in an unconscious way. In Bazzan et al. [1999] this layer is called the tactical layer; in Balmer et al. [2004] it's the physical layer. This layer connects the overall driver agent to the road network in which interaction with other drivers and also with the infrastructure are simulated. Local information about vehicles within a particular observation radius can be directly perceived and processed in the tactical layer, e.g., for safe driving regarding distances to leading vehicles. Information about surrounding traffic coming from the tactical layer or information about the state of

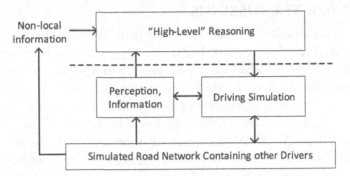

**Figure 6.1:** Advanced driver architecture pattern adding a high-level reasoning component on top of a micro- or mesoscopic traffic flow simulator. The reasoning component may influence driving parameters, the current route, or the destination.

traffic in other parts of the network (non-local traffic state obtained, e.g., from a radio broadcast, a variable message sign, or from direct information about blocked links) are input for the high-level reasoning layer. This layer contains the specific cognitive, high-level abilities represented in the advanced driver model. In Bazzan et al. [1999] this layer is called the "strategic" layer; in Balmer et al. [2004] it's the "mental" layer. Depending on the particular objective of the simulation, the explicit reasoning layer influences the parameters of the driving simulation (the desired speed, the acceptable headway distance, etc.). It can also determine the destination of the current trip, or the route that the simulated driver pursues. Next, we focus on different instances or applications of this general architectural pattern.

## 6.2.1 ANTICIPATION

One of the earliest extensions to the Nagel-Schreckenberg model, developed by Knospe et al. [2000], already aimed at including anticipation effects. Instead of simply adapting its speed to the gap to the vehicle ahead, a DVU pro-actively reduces its speed when it perceives that this vehicle is also executing some braking maneuver. This extension led to more realistic reproduction of traffic flow phenomena such as stop-and-go waves.

A similar idea of anticipation is present in ARCHISIM [Espie and Auberlet, 2007], where anticipation of the behavior of an agent was used to decide about maneuvers such as overtaking or braking in a freeway traffic simulation. The agents reason about the potential intentions of the vehicle ahead, such as whether or not to take the nearby off-ramp. Based on these anticipated intentions, the driver agent adapts its behavior. This enhanced model was validated and could produce simulation data that matched empirical observations better than traditional models without an anticipation component.

## 6.2.2   BEHAVIOR AT CROSSINGS

In microscopic traffic simulation, crossings are handled "centrally" by some form of controller that allows individual vehicles to pass. Vehicles then apply their usual driving model—eventually adapting their desired speed—when traveling into and through the crossing. The challenge here lies in un-signalized intersections, in which drivers potentially have to coordinate. Especially interesting scenarios are those in which, from time to time, a driver ignores priority rules. For this case, Doniec et al. [2008] introduce driver-agent coordination at an un-signalized intersection into the traffic simulator ARCHISIM [Espie and Auberlet, 2007]. When an agent approaches an intersection, it checks the priority laws such as "right before left" and maps them to the traffic situation. The agent also anticipates the behavior of the other agents at the intersection. On the one hand, the agent may yield to other entities that have priority. On the other hand, Doniec et al. introduce additional rules that describe violations of those laws. In a rule-based way, the agent reasons about whether its waiting time is too long ("impatience"), or if another driver is too slow. If these conditions are satisfied, the agent ignores the law and takes the chance to move across the intersection, although the law would not have allowed this. Thus, based on this norm-violating reasoning, accidents may occur. To avoid blocking at crossings, these authors added anticipation based on constraint propagation techniques.

## 6.2.3   EMOTION AND AGGRESSIVENESS

Simulating drivers with different personalities and how this affects their driving style is the aim of works that deal with high-fidelity driving simulators in virtual reality applications for making the traffic simulation more "life-like" (as for example in Wright et al. [2002]). As mentioned before, even in simple microscopic traffic flow simulation models, a driver's personality or driving style can be expressed by particular sets of parameters: an aggressive driver possesses a higher desired speed, a higher value for acceptable acceleration, and a shorter safety distance to leading vehicles than a more cautious driver.

A more elaborate psychological layer for agents in driving simulators is proposed in Seele et al. [2012]). This conceptual model is one of the major psychological personality models that is used for analysis of accidents in traffic psychology [Herzberg, 2009]. In Seele et al.'s model, the behavior that is not compliant with traffic rules (and with risky actions) is triggered by a personality-enhanced model. Following the "five factor model" each personality can be modeled by a combination of five distinct personality traits: openness, conscientiousness, extraversion, agreeableness, and neuroticism. Constellations of these five factors with different weights are used to compile three types of drivers: over-controlled drivers (closely following traffic rules), under-controlled drivers (more risk-taking behavior), and resilient drivers (mixed behavior). For example, an over-controlled driver is characterized by high neuroticism and high consciousness, average in agreeableness and low in openness and extraversion [Seele et al., 2012]. To illustrate the realism of the behavior of a group composed by a mixture of these three types, an un-signalized intersection is used as a scenario.

Also, from the point of view of traffic psychology, Maag et al. [2010] propose an emotional driver model for increasing the understanding of how emotions influence driving: driver agents perceive interactions with other agents (observing maneuvers such as lane changes or being blocked by a leading vehicle), which are associated with an "emotional surplus." In a cognitive-emotional layer of the architecture, the drivers' intention and personality influence how he interprets the current traffic situation. Based on this evaluation, the driver's emotional state changes, and influences parameters of the vehicle in a traditional microscopic traffic flow simulation. Maag et al. validate their concept based on "anger" observed with subjects in a driving simulator.

## 6.2.4  ROUTING, PLANNING AND BEYOND

Introducing the idea of a two-layered architecture for agents in traffic simulation, Bazzan et al. [1999] proposed a BDI architecture[1] for the strategic decision making. This high-level reasoning layer is responsible for route choice. Other approaches such as Rossetti et al. [2002] or Balmer et al. [2004] have added more reasoning abilities in the strategic layer. For instance, in addition to reasoning about the route choice, the agents may consciously reason about departure time and mode choice, thus involving also pre-trip information about the traffic state. To this aim, tools such as BDI-based agent specification and programming languages can be used as in Rossetti et al.'s work. Salvucci et al. [2001] go beyond folk psychology and explains how to use an established architecture from the cognitive science for modeling drivers' behavior. The architecture ACT-R (which is applied there) formulates a particular understanding about how humans reason.

Traditionally, re-routing of agents can be seen as a part of dynamic traffic assignment, as discussed in Section 4.3. Genuine agent-based approaches can be distinguished from those that just simulate traffic assignment procedures by the idea that the reasoning capabilities can be used at any time during the simulation. In case the simulated driver receives information that affects the links it plans to use, the agent is able to re-route immediately while still driving. This is also called en-route planning. Klügl and Rindsfüser [2011] illustrate that the overall load distribution is clearly different depending on the distance between the link where an incident has taken place and the location where re-routing is triggered. Other studies on re-routing can be found in Panwei and Dia [2006] using a fuzzy-neural architecture for deciding about changing routes, or Dia [2002] with a BDI architecture.

Combining reasoning components to form a complex agent architecture for traffic simulation has an essential advantage over traditional approaches, such as the four-stage model: all decisions related to demand modeling and mobility that a traveler has to make are integrated. Consequently, these decisions can be made in a consistent and rational way. It starts with the combination of route, mode, and also departure-time choice, and allows for the formulation of realistic adaptive behaviors, such as canceling a leisure shopping trip while stuck in a jam, so that an important date in the evening can still be met. By the inclusion of all choices in an integrated

---

[1]BDI stands for "Belief," "Desire," "Intention" denoting the three major components of the internal architecture of an agent. Although very popular, this architecture inspired by folk psychology is just one out of a set of possible reasoning architectures.

way, behavioral consistency can be guaranteed. For example, an agent cannot decide to take a car on its trip back home from work if it did not go by car in the first place [Nagel, 2007]. Therefore, it is not surprising that activity-based approaches integrating demand and mobility models have become very prominent with the advent of agent-based simulation in traffic.

## 6.3  TO KNOW MORE

Artificial Intelligence for traffic modeling and simulation is not restricted to enhanced models of individual decision making. In collections such as Bielli et al. [1994] or TRB [2007], AI techniques are proposed for processing data for use in simulation validation. Decision making being influenced by social behavior is the subject of Illenberger [2012] or Ronald et al. [2009].

For introductory material about BDI, the reader is referred to a paper by Rao and Georgeff [1991], as well to a textbook by Wooldridge [2009].

The reader interested in the seminal papers about car-following and lane-changing models are referred to Gipps [1981] and Gipps [1986]. Lane-changing models in microscopic traffic simulation may result from a conscious decision making as in Luo and Bölöni [2012].

# CHAPTER 7

# Intelligent Measures in Control and Management

As seen in previous chapters, there are several strategies of traffic control and management, which may relate to one or more of the following elements: the physical network, how this network is modeled, control devices installed in this network, the demand, and disturbances of several natures (that can be measured, detected, or forecast). Some of these are computational entities (e.g., models, surveillance, and control strategies). The goal of the present chapter is to discuss these entities in more detail, as they have increasingly been the focus of computer science and AI researchers. We start with techniques that address traffic signal control as this was traditionally the focus of the majority of works arising from the computer science community. This appear in the next section. In Section 7.2 we discuss approaches that go beyond classical control using traffic signals. For instance, we address approaches that combine classical strategies with control of the demand, and with intelligent routing. Also, we discuss strategies that deal with un-signalized intersections.

## 7.1 STRATEGIES FOR INTELLIGENT TRAFFIC SIGNAL CONTROL

The strengths of progressive systems (i.e., coordination of traffic signals) such as TRANSYT, SCOOT, SCATS, and similar systems were discussed in Chapter 2. However, as seen in Section 2.4.2, these methods also have some drawbacks. Some of these drawbacks have motivated computer science and AI researchers to come up with intelligent computational solutions. In particular, it is worth mentioning two of these drawbacks: the assumption of "typical" conditions (seldom found in the real-world) and the *a priori* determination of the appropriate signal plans for the different times of the day (a complex task that requires a lot of domain knowledge).

Several researchers have addressed these issues by designing more intelligent strategies for traffic signals control. These strategies address a range of issues that goes from isolated controllers to a network of controllers, in which a progressive system implements green waves in more than one traffic direction. In Section 7.1.1 we present some of these approaches. Due to the importance of learning-based systems, we discuss these separately in Section 7.1.2.

## 7.1.1   FROM ISOLATED TO COORDINATED INTERSECTIONS

Regarding single controllers, typical works are those from Kosonen [2003] and Koźlak et al. [2008]. In the former, various agents interact in a single intersection using fuzzy inference. Each phase is modeled as an agent; agents then negotiate to decide when to change the lights of the group to green. Koźlak et al. [2008] proposed a multiagent system to control traffic signals based on intersection managers (called intersection coordinators). Besides these managers, agents are also in charge of predicting future traffic conditions.

A more complex family of optimization problems deals with networks of agents located at intersections. This poses extra problems because agents may need to coordinate with neighbor agents. There is no obvious protocol that prescribes how the non-local information should be integrated with the local one. Some popular approaches are: (i) organizing control agents in a hierarchical structure so that conflicts are solved at an upper level; (ii) letting agents learn how to control their local environments or coordinate their actions; and (iii) let agents self-organize. In the particular case of learning-based approaches, unless agents receive utilities for their actions that are aligned with the global objectives of the system, the performance at global level tends to be poor.

As seen in Section 2.4.2, one difficulty in extending the synchronization to a network, or to more than one direction of traffic, is the fact that in some key intersections conflicts may appear because different directions compete for bandwidth. One approach is to let a traffic expert solve these conflicts. Alternatives to this approach seek to replace the traditional arterial green wave by shorter green waves in segments of the network. This can be done, e.g., using negotiation over the question of which traffic direction shall be given more bandwidth.

To do so, an approach based on distributed constraint optimization problems (DCOP) was proposed in Oliveira et al. [2005]. It is intended to be a compromise between the totally autonomous coordination (with implicit communication) that is described in Gershenson [2007], and the classical centralized solution (e.g., TRANSYT). The constraints in this problem arise from the fact that, in each intersection, a traffic signal cannot coordinate with all neighbors at the same time. A conflict occurs when two neighbors want to coordinate in two different traffic directions. Junges and Bazzan (2008) later extended this scenario to bigger networks, aiming at investigating computational issues related to performance, such as time to reach an agreement and number of exchanged messages.

In order to improve the solution provided by the TUC approach (Section 2.4.2), predictive control is used by de Oliveira and Camponogara [2010]. The optimization problem arising from the model predictive control is solved by decomposing it into a network of coupled, small sub-problems to be solved by agents located at intersections. These agents sense the state variables and set the values of the control variables, communicating with agents in the neighborhood in order to acquire data from these and coordinate their actions.

An alternative to using explicit communication among all agents is to use some kind of organizational structure, where a hierarchy of authority exists. Roozemond [2001] proposed an agent-based urban signal control system that consists of several intersection control agents, authority agents, and road segment agents. Authority agents supervise and control several intersection agents, each of these in charge of managing one intersection. Neighboring intersection agents also communicate. Intersection agents use a prediction model that estimates the traffic situation, as well as predetermined rules to select a control strategy. Based on these predictions and on a set of rules, the intersection agent computes signal plan timings, check with adjoining agents, and plans the signal control strategy. This is sufficient to achieve good results in non-conflicting cases. The conflicting ones are passed to the authority agents. The author mentions that an intersection agent must often sacrifice some performance in favor of cooperative behavior, and that some sort of negotiation is necessary. Given that conflicts must be solved in real-time, this is likely to be key.

Another hierarchical multiagent system is proposed by France and Ghorbani [2003]. In the first level, local traffic agents (LTAs) represent intersections; these are responsible for providing appropriate traffic signal timing. In the second level, a coordinator traffic agent (CTA) supervises a few LTAs.

One way to achieve progressive systems is to let agents self-organize regarding the decision about which green waves form. Methods that do not use communication between traffic signals were especially popular until some years ago, when the cost of communication (e.g., wireless) was still prohibitive. For example, de Oliveira et al. [2004] proposed an approach based on swarm intelligence. Each intersection controller behaves like a social insect. Signal plans are seen as tasks. Stimuli to perform a task or, sometimes, to change tasks, are provided by the vehicles that continuously produce "pheromone" while waiting for green lights. Thus the volume of traffic coming from one direction can be evaluated by the intersection agent, and this may trigger some signal plan switching. Experiments have shown that the agents achieve synchronization without any central management. However, the time needed to converge to a stable coordination can be high, which is a negative aspect especially in highly dynamic environments.

A similar idea appears in Helbing et al. [2005], where vehicles can synchronize traffic signals and organize green waves. In Gershenson [2007], traffic signals self-organize by means of three methods, also with no direct communication between them. It is shown that the adaptation to traffic conditions reduces waiting times and the number of stopped vehicles.

Bazzan [2005] used techniques of evolutionary game theory and stochastic games, in which there is no explicit communication among agents: traffic signal agents act in a dynamic environment having only local knowledge. These agents perform experimentation and receive a reward that depends also on the experimentation performed elsewhere in the neighborhood. Stochastic events that may take place in the network are modeled by mutations. During the learning process, a fitness for each strategy is computed and influences the next generation of strategies that is used by one agent. A shortcoming of the approach is that payoff matrices have to be explicitly formalized by the designer of the system. This makes the approach time-consuming when many different options of coordination are possible. Thus, a challenge here is how to design mechanisms for large-scale of networks.

In Prothmann et al. [2011], an observer/controller architecture for signal control is proposed. This effort appears within the framework of organic computing. The idea is that a signal controller is the system under investigation, while an observer monitors the local traffic demand and evaluates the performance of the active signal plan. Based on the observations, the controller selects and optimizes signal plans by means of a two-leveled learning mechanism. In level 1 signal plans are selected on-line from a previously learned mapping. This relies on a variant of a learning classifier system. Level 2 is in charge of optimizing the available signal plans when an unknown demand arises. This optimization is achieved by means of evolutionary algorithms.

## 7.1.2   APPROACHES BASED ON REINFORCEMENT LEARNING

A popular approach related to control of traffic signals is to model these signals as agents whose goal is to learn a policy for mapping states to actions by means of reinforcement learning and Markov decision processes (MDP). Depending on the actual formulation of the MDP, it may not be computationally feasible to solve the problem because the space of state-action pairs grows exponentially. This is due either to the discretization regarding the number of states in a single intersection, or to the number of intersections, or both. Solutions for this problem have been proposed in two directions. First, instead of modeling the whole problem of $n$ intersections in a single MDP, one may use $n$ independent learners. Second, it is possible to use function approximation to deal with the size of the space state. The former is a popular approach because it has the advantage of scaling to a large number of agents. However because agents learn independently in an environment where other agents are doing the same, this approach may lead to sub-optimality.

Next, some of the proposed approaches are reviewed; we start with those based on independent learners (for network of traffic signals as well as for single intersections), and later move to approaches based on some kind of interaction or coordination between the agents that represent intersections in a traffic network. While in the latter case the interaction between agents is explicit, in the case of independent learners interactions do happen but they are just implicit.

### Independent Learners

One of the first works dealing with reinforcement learning for traffic control was based on stochastic games [Camponogara and Kraus Jr., 2003]. The emphasis is on applying one of the following policies: uniformly random policy (assigns the same probability to all actions available to an agent); best-effort policy (lets traffic flow from the lane with the longest queue); and one achieved by using Q-learning. For the experiments, two intersection agents were used. Depending on the traffic conditions, if both agents implement Q-learning, there is a significant reduction in the waiting time.

Q-learning is a model-free approach to reinforcement learning. This means that the agents do not start with any full or partial model (policy) that maps states to actions. It has the advantage of being simple, but it may not perform well in non-stationary environments. Therefore, in Silva et al. [2006], the authors propose the use of a model-based approach, in order to deal with non-stationary nature of traffic flow patterns. In this approach it is assumed that flow patterns are non-stationary but they can be nearly divided into stationary dynamics that need not be known *a priori*. Each model is assigned to an optimal policy (which is a mapping from traffic patterns to signal plans), and to an estimate of its quality. The creation of new models is controlled by a continuous evaluation of the prediction errors generated by each partial model. If the environment changes and a local policy turns suboptimal (congestion increases over a threshold), then the system creates a new model. Whenever possible, the system reuses existing models instead of creating new ones.

In Prashanth and Bhatnagar [2011], the learning problem is solved in a centralized way with a single MDP. Due to the size of the space of pairs state-action, their scenario is restricted to a grid of $3 \times 3$ controllers. Clearly, the centralized MDP poses a limitation for real-world applications. On the other hand, the authors focus on an important problem, namely the high dimension of the space state, by using a function approximation technique.

**Interacting Learning Agents**

Given the computation complexity of a centralized MDP vs. the sub-optimality of independent learning, a recent trend is to let controllers learn independently but allow them to interact and combine their policies. This way one may depart from total centralization, as well as from total independence, and find a compromise between these two extremes. We now discuss some of these approaches. These are mostly based on agents being able to coordinate their actions by means of several techniques, such as communication of advice, collaboration, mediated learning, etc.

The question of how heterogeneous groups of agents can benefit from communication to improve their learning skills was investigated by Nunes and Oliveira [2004], in which information from several sources is used in a simplified traffic control problem. Members of a team of agents may communicate among themselves or with members of other teams that are solving similar problems in different areas. Different types of agents use various techniques, such as neural networks and heuristics.

A similar, collaborative approach is proposed in Salkham et al. [2008], with the difference that only reinforcement learning techniques are used. The authors call this approach collaborative reinforcement learning (CRL). Each intersection has a CRL-based agent to control the traffic signal. These agents collaborate with neighboring agents in order to learn appropriate phase timing based on the traffic pattern. The approach was tested using Dublin's inner city center. The collaboration is governed by a specific advertisement strategy. This strategy defines the frequency of communication, the nature of communicated data, and the groups of CRL agents every controller is allowed to send to and receive from. This paper focuses on agents employing a common advertisement strategy that allows them to exchange their rewards.

The work of Kuyer et al. [2008] also focuses on cooperative learning and explicit coordination among agents, i.e., joint actions among them are considered. However, this leads to an increase in complexity when compared to independent learning. Thus, some simplifying assumptions must be made. For instance, it is assumed that the coordination takes place only among direct neighbors in the network. Under this assumption, the global coordination problem may be decomposed into a set of local coordination problems and can be solved with the use of coordination graphs as proposed by Guestrin et al. [2002]. To find the optimal joint actions in such graphs, the authors apply an algorithm that estimates the optimal joint action. This work assumes that relationships among controllers are restricted to their direct neighborhoods. However, in traffic this is not always the case. Although coordination graphs can be extended to other kinds of neighborhoods, this approach has problems handling a high number of agents in each graph.

Exchange of information by independent controllers is also tackled by Oliveira and Bazzan [2009]. Thus, this work analyses how agents can benefit from sharing information as well as the consequences of this cooperation in the performance of the traffic system. The authors show that having information is not sufficient if the other agents are not acting in a coordinated way; extra information can have a negative effect in the learning process if the information is unnecessary. Also, noise in the traffic patterns and the inter-dependencies among the agents' actions are relevant factors to consider when there is a need to decide between using shared information, using only local information, or an in-between solution.

One approach to reduce the complexity in large networks of traffic signal controllers is to explore their organization. In Bazzan et al. [2010], a layered architecture is proposed, where supervisor (or mediator) agents are in charge of a small group of controller agents. This tries to balance the number of joint vs. independent actions using a kind of organizational control in line with the work by Zhang et al. [2009].

Another work in this direction is by Abdoos et al. [2013], which proposes the holonic multi-agent system (HMAS) to model a large traffic network that is partitioned into a number of regions; holons are then assigned to control each region. The holons are hierarchically arranged in two levels: intersection controller holons (these can be seen as the controller agents proposed in the previous works discussed here) in the first level, and region controller holons in the second level. Q-learning is then extended to control the signals in both levels.

Organization of agents in regions is also explored in Choy et al. [2003], although the focus of this work is on using a fuzzy-neuroevolutionary hybrid system with online reinforcement learning for control of a network of traffic signals. The main idea is to overcome weaknesses of neural networks regarding reasoning and inference, by combining them with fuzzy systems for knowledge representation. The paper investigates the use of two methods. The first is online reinforcement learning facilitated by fuzzy logic. Here, reinforcements are derived from the fuzzy estimation of state changes between time intervals. Weights adjustment of the fuzzy-neural systems are then carried out by back-propagation of the reinforcement. The second method is to perform testing using hypothetical trial solutions on a model of the system. Better solutions are subsequently generated using evolutionary techniques based on the suitability of performance of the trials. The hypothetical solutions represent the fuzzy relation between the antecedents and implication of the fuzzy rules. Besides this hybrid method, the organization of the network is also explored: a large traffic network is divided into regions; each region is further subdivided into zones, which are made up of several intersections. Similarly, there are three types of hybrid agents: intersection controller agent, zone controller agent, and region controller agent. Agents at the lowest level (intersection controller agents) control pre-assigned traffic intersections while agents at the higher level such as the zone or regional controller agents manage several lower level agents. Each agent acquires fuzzy knowledge of the traffic situation in a different manner depending on its hierarchical level and makes autonomous decisions to control the traffic. In addition, the element of cooperation is embedded within each decision. Individual agents decide whether cooperation is needed for the controlled area.

## 7.2    BEYOND PURE TRAFFIC SIGNAL CONTROL

### 7.2.1    APPROACHES EXPLICITLY ADDRESSING DEMAND

So far, we have discussed traffic control approaches that focus on signal optimization. In the following, we discuss works aiming at integrating both the demand and the control sides.

Some approaches for control at intersections propose that the waiting time of drivers is considered when computing or deciding the timing of traffic signals. In this line Wiering [2000], describes the use of reinforcement learning by traffic signal agents in order to minimize the overall waiting time of vehicles in a small grid. Agents learn a value function that estimates the expected waiting times of vehicles given different settings of traffic signals. One interesting issue tacked in this research is that a kind of co-learning is considered: value functions are learned not only by the traffic signals, but also by the vehicles which can thus compute policies to select optimal routes to their respective destinations. This vehicle-based representation enables the estimations of reward values.

A similar RL-based method for controlling traffic signals is presented by Steingröver et al. [2005] to minimize the total travel time of all vehicles in the network. Thus, the control perspective is a global one, although the actions are local. Agents are the traffic signals but the learning task is formulated in a way that the state representation is vehicle-based (waiting times for individual vehicles), aggregated over all vehicles around the intersection. In their decision making processes, traffic signals consider information about the vehicles. The paper also investigates more efficient forms of state representation, with different learning abilities, generalization, and performances.

In Bazzan et al. [2008], the learning task also involves two kinds of agents: drivers and traffic signals, each having its own goal and learning algorithm. The objective of local traffic control is to minimize queues in a spatially limited area (e.g., around a traffic signal). The objective of drivers is to minimize their travel times experimenting with several routes, one in each commuting episode. Later, the authors have extended this method to incorporate on-the-fly re-routing [Bazzan and Klügl, 2008]. Here drivers react to their perception of jammed links and adapt. This means that new routes are not necessarily learned.

## 7.2.2 COORDINATION OF DRIVERS' CHOICES

Tumer and colleagues tackle congestion problems in a broad way using the metaphor of minority games. In Tumer et al. [2008], multiagent learning algorithms are applied to two formulations of the problem of selecting departure times. These formulations are bottom-up and top-down. Different time slots have different desirability that reflect users' preferences. The system utility is measured from the perspective of a city manager that seeks to minimize system-wide delays. From the perspective of drivers, these aim at maximizing a personal objective function (e.g., the difference between the desired and the actual arrival time). In both cases, the fact that agents greedily pursue their best interest causes traffic to worsen for everyone. Agents' actions are determined based on a reinforcement learning algorithm. The key issue in this work is to ensure that the agents receive utilities that promote good systems level behavior.

In [Tumer et al., 2009] a similar approach is used to explore the impacts of agent reward functions on two traffic problems: selection of departure time and selection of lane. The authors make an important remark about one issue that arises in traffic problems but does not arise in many other domains (e.g., rover coordination), namely ensuring that drivers follow the advice they receive. A related problem also arises when the city manager's reward is at odds with a social welfare function. Determining what incentives to provide to the agents so that these two seemingly different objectives can be simultaneously maximized is a critical problem that bears further study. This interesting approach is perhaps more suitable to a scenario composed by autonomous vehicles because in this case rewards could be communicated. Human drivers do not have many means of perceiving reward, apart from, for instance travel time. However, travel time alone is not necessarily a good measure of global utility. Thus, a challenge here is to design appropriate mechanisms that ensure that drivers are getting utilities that are both aligned with the system reward, and are as sensitive as possible to changes in the reward of each agent.

## 7.2.3 LIGHTLESS AND MARKET-BASED APPROACHES

Mandiau et al. (2008) consider a single intersection, without traffic signals, having only stop signs as a two-player game played by autonomous vehicles approaching or entering the intersection. The actions available for the autonomous vehicles are to stop or to move.

The seminal work by Dresner and Stone [2004] has proposed a reservation-based intersection control where autonomous vehicles try to cross an intersection without conventional traffic signals. It was assumed that: autonomous vehicles are not allowed to turn, do not change lanes, and all begin traveling roughly at the same speed. The reservation is performed as follows. First, each autonomous vehicle informs the intersection manager (IM): the time it will arrive at the intersection, the velocity, direction, maximum and minimum acceleration and other vehicles properties. Then, the IM simulates the journey of the autonomous vehicle given the IM's knowledge about other similar reservations. If the space requested by the autonomous vehicle (for a given timeslot) is already occupied, then the request is rejected, in which case the autonomous vehicle must decelerate and try again. If the request is accepted, it must be kept or canceled by the autonomous vehicle (in case it cannot be met).

Later, in [Dresner and Stone, 2008], an improved protocol was proposed, based on some rules that the vehicles are expected to follow. Also, a modification in the protocol allows autonomous vehicles to share the system with conventional vehicles.

Finally, in Au et al. [2011], a further management policy is proposed, which puts the request messages on hold and then process several requests at once, in order to avoid starvation at secondary roads.

The reservation-based approach was also extended in Dresner and Stone [2006] to address deadlocks and delays caused by drivers putting reservations in a suboptimal way because the intersection manager processes requests on a first come, first served basis.

This has also inspired the work of Vasirani and Ossowski [2009, 2011], where the reservation-based approach is extended to cover networks of intersections. Drivers trade with the infrastructure agents in a virtual marketplace, purchasing reservations to cross intersections. Market rules were designed with the aim of aligning the global profit (revenues from the infrastructure use) with the social welfare (e.g., average travel time), in a way that, in situations of similar traffic load, an increase of the infrastructures monetary benefits usually implies a decrease of the drivers average travel times.

Schepperle and Böhm [2007, 2009] proposed an approach that takes the valuation of the drivers into account. The focus is on single intersections. In Schepperle and Böhm [2007] a procedure is proposed, which consists of four steps: vehicle contacts the intersection; vehicle acquires an initial time slot to cross the intersection; if not satisfied, a vehicle can try to acquire a better time slot, this time from another vehicle; vehicles cross the intersection. In the second step, an auction (e.g., a second-price, sealed-bid) is run among the vehicles that do not yet possess a time slot. A variant is proposed in which a vehicle with a high valuation subsidize another vehicle that is located ahead and that is going to bid for a time slot but has a low valuation. Subsidizing vehicles ahead is expected to guarantee a better time slot for both vehicles. In the third step vehicles arriving late can acquire time slots that have already been auctioned off. In Schepperle and Böhm [2009], the authors explicitly discuss some challenges for making valuation-aware control systems operational.

Tavares and Bazzan [2012] present an approach that considers adaptation by both the infrastructure (link managers) and the drivers. While a link managers update link prices to cope with the varying demand, drivers try to adapt to the road network changes in order to minimize their costs. In their case, some drivers care about travel time, while others prefer to spend less.

Balan and Luke (2006) proposed history-based controllers intended to provide a kind of global fairness. They base their approach on the notion of historical fairness by allowing vehicles to store credits they receive when waiting at red lights, and cash the credits in when passing through intersections. Traffic signals base their decisions on the credits of various vehicles at the intersection. When a vehicle reaches its destination, it reports its average waiting time over all intersections and this is one metric used to assess the efficiency of the control.

## 7.2.4  SYSTEM-LEVEL MANAGEMENT

Traffic management is another domain where there are opportunities for cooperating agents. This is probably due to the fact that agent-related technologies facilitate modularity, modeling, and simulation of management strategies. In this line, many tools, environments, and prototypes have been proposed. Apart from those already discussed in Chapter 6, next we present some propositions for developing integrated environments for traffic simulation and control, in order to better manage the traffic system as a whole.

In van Katwijk and van Koningsbruggen [2002], concepts related to agents and multiagent systems were used to formulate examples of traffic management based especially on heterogeneous agents and system components. The BDI agent architecture was suggested to represent and reason about agents' knowledge. Also, a coordination protocol based on priority over resource allocation is discussed that helps the agents to distribute the flow of vehicles. Resources that agents can manipulate are mainly the flow of vehicles and allowed speed. The authors mention some kinds of resource-bounded decisions agents must make: amount of resource to request, time of the request, which priority the request has, contingency for non-granted requests and/or changes in priority order. The authors discuss the problems that are ultimately tied to conflict resolution and the problem of local versus global performance. They try to solve the latter by formulating a hierarchical organization where authority both solves conflicts and pursue global performance. These issues were tackled in the recent literature but remain partially unsolved. An obvious challenge here is how to solve conflicts that appear due to insufficient resources and about which model of organization to adopt, namely a hierarchical, authority-based one, versus a totally decentralized with lesser commitment with global performance.

In van Katwijk et al. [2005] a testbed for traffic management is presented, which intends to manage different levels of complexity, a diversity of policy goals, and different forms of traffic problems. Their testbed aims at rapid development of multiagent-based management and control, consisting of an interaction model, intelligence models, and a world model. The former aims at modeling interactions among the agents, mainly via communication. Intelligence models are supported by rule-based inference, and by Bayesian inference (to model uncertainty agents may have). The authors claim that these three components are useful to implement decentralized traffic concepts that were reported in the literature. However, a closer analysis seems to indicate that their work is particularly suitable for hierarchical organized controllers and for inter-controller coordination. Therefore, an evident challenge is to extend this testbed to accommodate other kinds of organizations of controllers, intra-controller coordination (e.g., among phases or groups of phases that exist in an intersection), as well as other forms of coordination such as those not explicitly based on communication.

In Wang [2008] an approach is presented as a generalization of the feedback control mechanism in control theory. Wang's approach (the TransWorld) is based on a connection between the actual transportation system and its artificial counterpart. This connection is manifold, i.e., has many modes. In the learning and training mode both the artificial and the actual systems are only loosely connected and the former serves as data center for learning operational procedures. In the experimentation and evaluation mode, the artificial system serves as a platform to conduct experiments and eventually predict the behavior of the actual system. In the control and management mode, which is the most challenging one, both systems must be tightly connected (for instance, real-time connected). It is expected that the artificial system is able to replicate the actual behavior. Eventual differences can be used to generate feedback control. The author emphasizes the need for parallel execution of both systems, and that the artificial one is seen as a generalization of adaptive control whereas the AI-based artificial system replaces an analytic reference model.

The underlying idea of artificial transportation systems is also present in the MAS-T$^2$er Lab project [Rossetti et al., 2008], which is based on agent programming that can deal with the social and behavioral models.

In Chen et al. [2012], the authors argue that there is a need for dynamic selection of the most appropriate agent for each specific traffic state, and outline a method for recommendation of a control agent for a given transportation system.

Regarding the management of systems that include or are based on autonomous vehicles, in Bazzan et al. [2012], an agent- and market-based approach is proposed to manage a fleet of automated guided personal rapid transit vehicles (e.g., pods). This study considers some variants for both the processing of demands (trips) and for the routing. These variants are centralized versus decentralized, with or without en-route re-planning, and action based vs. first in, first served basis. Results show, at least in some cases, the feasibility of a centralized service. However, communication as well as reliability and fault-tolerance are, of course, important issues.

## 7.3    TO KNOW MORE

Traffic signal control is a popular research topic thus this chapter could not present but a small fraction of the work proposed and developed so far. There are two relatively recent surveys that cover works agent-based approaches in traffic and transportation problems. While the survey by Chen and Cheng [2010] covers a broader spectrum (e.g., railway transportation and pedestrian simulation), Bazzan and Klügl [2013] focus on vehicular traffic, discussing more deeply some agent-based approaches to modeling, simulation, control and management, thus addressing both supply and demand. Specifically for learning-based approaches (mostly control perspective), the reader may find further references and discussions in Bazzan [2009].

Other control measures exist, which focus not only on urban traffic control, but also on freeways. These include ramp control, high-occupancy lanes, and variable message signs (see Gordon and Tighe [2005]). Besides, congestion pricing can be used to control traffic. See Tsekeris and Voß [2009] for a review.

Apart from approaches that are based on autonomous agents and multiagent systems techniques, there is a broad literature published in IEEE journals as well as in publications from the area of control and computational intelligence, especially for works that combine AI techniques and fuzzy control. The reader may also find further pointers in journal such as *Journal of Intelligent Transportation Systems, IEEE Transactions on Intelligent Transportation Systems,* and *Transportation Research* (especially the TR-C).

# CHAPTER 8

# Driver Support and Guidance

In Chapter 4, we discussed assignment and route choice mostly from a system point of view. We also introduced traffic system analysis from the driver perspective, but nevertheless, the aim was the analysis of the overall traffic system. In this chapter we take the driver's perspective, dealing with support to the driver. Here, the overall system perspective is only insofar interesting as it may trigger a support system to give a particular advice. Thus, in the present chapter we address which intelligent approaches can be used for helping the driver to find the "best" route or the "best" location, and how relevant information can be provided to the driver.

We approach the overall topic on three levels: the actual driving level with advanced driver assistance, then the routing level with in-vehicle route guidance systems, and then a third level, with pro-active, opportunistic routing to recommended destinations.

## 8.1 (ADVANCED) DRIVER ASSISTANCE SYSTEMS

Driver assistance systems (DASs) and advanced driver assistance systems (ADASs) aim at supporting the actual driving task in order to increase safety and comfort for the driver, as well as to increase traffic flow by avoiding accidents, and to increase the capacity of the roads by decreasing the necessary distance between vehicles. We remark that there is no standard terminology when referring to these new technologies related to ITS. DASs and AVCSS (Advanced Vehicle Control and Safety Systems, discussed briefly in Section 1.3) are related. DAS seems to be a more general term. In this chapter we use the term DAS, as it emphasizes the support to the driver rather than the control of the vehicle.

Different subsystems are commercially available: (i) lane departure warning and lane keeping systems; (ii) adaptive cruise control and stop-and-go control in congestions; (iii) pedestrian protection; (iv) collision warning or avoidance; (v) parking assistants; (vi) automated stabilization; and (vii) automated wipers connected to rain and other sensors on the windscreen. Some of these sub-systems can be traced back to robotics and control and thus combine research on control systems with AI.

Driver support tools differ a great amount in complexity, ranging from simple control loops to sophisticated modules that may fully automate driving, such as a parking assistant. They show different levels of autonomy, giving warning to the driver, or working without human intervention. In the following, we discuss the components that make up an ADAS, followed by some overview to newer trends towards cooperative driver assistance system. In the next chapter we will elaborate more on automated driving as one of the trends that is expected to govern future mobility.

## 8.1.1  ELEMENTS OF A DRIVER ASSISTANCE SYSTEM

One important objective behind ADASs is to increase safety by avoiding accidents due to driver mistakes. Umemura [2004] distinguishes between three phases of driving in which mistakes may happen:

- Perception: the driver observes and categorizes a situation in her surroundings, containing environmental objects, but also involving the state of the vehicle itself.

- Assessment: during this phase the driver reasons about perceptions from the environment, anticipates what may happen, and makes a decision about what to do. In the situation aware-ness literature [Papp, 2012], this phase can be further divided into "comprehension" (for the integration of data and the determination of its relevance for the particular objectives) and "projection," which involves dynamics of observed phenomena and the anticipation of fu-ture events.

- Operation: the driver actually does something, e.g., release the accelerator.

These three phases correspond to the classical "perceive-reason-act" cycle of agent-based approaches in AI. The words originally used in [Umemura, 2004] are "cognition" for perception, and "judgment" for assessment.

DASs may give support to all three phases, but do not necessary need to cover all [Röckl et al., 2007]. Figure 8.1 illustrates the overall architecture starting from the three phases of driving.

A DAS needs to perform steps that are similar to those performed by a human driver (e.g., those that allow the observation of the situation around). Typically, radar sensors are used for obstacle detection, due to their independence from day or night time, as well as high tolerance to imperfect weather conditions. An alternative is the use of laser sensors, due to their higher angle of sensable area. Vision-based systems are mostly used for control of the car's lateral, yet they are more expensive as they need image processing capabilities [Papp, 2012]. Sensor information is evaluated and integrated to create the necessary level of situation awareness for the actual service, e.g., the distance to the lane margins or whether there is an obstacle ahead, etc. Overall situation awareness also includes information on the state of the driver. Thus, in addition to environmental context there should be a component for monitoring and anticipating the driver's reaction in the critical situation.

In an information fusion step, the environmental context and the driver context (e.g., the driver's state in terms of tiredness) are integrated and the appropriate reaction is derived. The overall control loop is closed either by warning the driver (e.g., by an acoustic signal or vibration), or by actually intervening (e.g., by braking or avoiding).

The process of handling sensor information can be done separately for each sub-system. Yet, a more efficient solution can be achieved by integrating the different assessment results into a larger data set representing the full context of the driver. Clearly, the more complex the control loop, the more extensive the situation assessment needs to be.

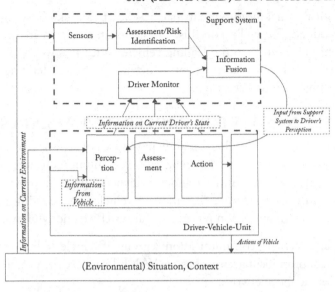

**Figure 8.1:** Overall architecture of the driver-vehicle-unit plus driver support system (Umemura, 2004, Röckl et al., 2007).

Various techniques and approaches have been applied and suggested for the different components of a DAS. Many of those "autonomous"[1] systems are developed by car manufacturers for increasing the attractiveness of their products. Due to this commercial success, a variety of publications is available surveying ADASs from different perspectives. An important trend leads to "intelligent vehicles" that are equipped with combinations of these support systems for increasing safety, comfort, and, ultimately, the traffic flow. Eskandarian [2012] surveys an extensive collection of relevant information for ADASs. Röckl et al. [2007] suggest a general architecture for a DAS and the usage of Bayesian networks for information fusion of uncertain information.

Designed to increase the safety and comfort of drivers, smart driving support tools have an impact on driving performance. In the beginning of automotive support systems, research and development focused on providing the driver with as much information as possible. Consequently, information overload became a topic in designing ADASs. Meanwhile, one can find a large volume of research in traffic psychology and ergonomics for analyzing the effect of different types of ADASs on the driver and her behavior, her feeling of security, distraction, etc. From the beginning, behavioral response was an important topic [Brackstone and McDonald, 2000]. Systems following the idea of "driver-in-the-loop" were promoted, for example in IHRA [2010] and Stevens et al. [2002]. Hajek et al. [2013] even propose an adaptive cruise control system

---

[1]Papp [2012] calls these purely sensor-based systems "autonomous," as all relevant information is collected by the controlled vehicle itself. This is in contrast to cooperative systems, in which vehicle-to-vehicle and vehicle-to-infrastructure communication enhances the information sources.

that uses physiological signal measurements (such as heart-rate, skin response, or respiration) for detecting high workload. Depending on this, a cruise control system adapts the configuration of the acceptable minimal distance to the leading vehicle. With this work, they could show that it is technically possible to build such a system, which would be—at least according to subjective evaluation by test subjects—well accepted.

## 8.1.2 NEW TECHNOLOGY FOR COOPERATIVE ASSISTANCE SYSTEMS

Advances in communication technology, mainly in the area of ad-hoc networks, enable so-called cooperative ADASs as the next step after what is called autonomous assistance systems. In the last two decades, there has been massive funding given to projects providing reliable and fast communication technology for connecting moving vehicles and moving vehicles to stationary infrastructure. "V2X" communication has two instances [Piao and McDonald, 2008].

- V2V (Vehicle to Vehicle) communication: a group of vehicles is linked and enabled to communicate. Most prominent approaches are VANETs, where the vehicles are nodes in a temporary wireless ad-hoc network.

- V2I (Vehicle to Infrastructure) communication: vehicles communicate to special infrastructure modules. Through this, an operator can interact directly with the individual vehicle (1:1 communication), or broadcast information to all passing vehicles that are equipped with compatible devices (1:many). Communication can be one-way or bi-directional, enabling information exchange.

Kakkasageri and Manvi [2013] give a review (from a technology-oriented point of view) on information transfer in ad-hoc networks created from vehicles. V2X allows the distribution, to equipped vehicles, of information that is not accessible to their restricted, local sensors. This information sharing may have the following effects. First, up-to-date information on relevant context of driving can be shared by infrastructure modules. A traffic operator may directly broadcast information about critical weather conditions on a fine spatial and temporal resolution. Second, a vehicle may share data from sensors that it is equipped with, yet others in the network do not have. Also, hazards ahead can be communicated to vehicles behind, in order to prevent multi-car accidents. This is especially useful in situations with reduced visibility, incidents outside the range of the sensors. Third, automated direct information transfer may distribute fast information on acceleration actions or emergency braking of the leading vehicles. It is expected that this will enable smaller safety distances or even platoon-like driving by bridging delays on human reaction time.

These effects of information sharing and transfer promise to enhance safety and comfort, as well as to improve the overall traffic flow by the enhanced situation awareness provided by communicating support systems. Yet, whether this really happens is unclear. Piao and McDonald [2008] compare the potential effects of using collaborative or autonomous assistance systems in cruise control to collision avoidance. First, they discuss the promised positive impact of adaptive cruise control assistants on traffic flow. The basic objective behind this system is to increase driver comfort by allowing the driver to delegate extenuating speed-brake actions to the system. However, because of this objective, the maximum allowed acceleration and deceleration has to be limited. Consequently, the driver must have the chance to intervene when an emergency brake action is necessary. Therefore, the minimal acceptable distance that the assistant is allowed to maintain is almost the same (or even slightly larger) than the one the driver would have chosen without the assistance system. Thus, in a standard traffic situation, one should not expect a positive impact on the road capacity.

In contrast to this, sharing information on the level of routing may have positive effects: Yang et al. [2010] show that information exchange among vehicles may help to improve their distribution in the traffic network, thus reducing travel times.

Some research works propose coordination of vehicles organized in platoons as, e.g., Desjardins et al. [2009], where the authors describe an agent-based cooperative architecture that aims at controlling and coordinating vehicles joining platoons, traveling in a platoon, and leaving the platoon again. This is achieved (in a simulated scenario) using a multi-layered architecture, with an action layer for low-level vehicle control actions such as braking, accelerating, or steering and a coordination layer which is responsible for high-level action choice by integrating cooperative decision making between vehicles. In both layers, multiagent reinforcement learning techniques are used, showing that the integration of these techniques at all levels of the autonomous driving controller yields efficient results for vehicle control and coordination.

The case of platoons of heavy-weight freight vehicles is different. There have been successful tests of (heterogeneous) trucks, which were virtually connected into a platoon with minimal distances between them [Chan et al., 2012].[2] One can expect that such "Road Train" systems will be one of the major advances for increasing infrastructure capacity.

Nevertheless, cooperative ADASs are still in an early phase and far from practical use in private cars. Open questions are which kind of protocols can be used and how they related to communication efficiency and robustness (see, e.g., Lochert et al., 2008, Rybicki, 2011, Scheuermann et al., 2009).

Discussing cooperative collision warning and avoidance assistants, Piao and McDonald [2008] also raises another critical issue: in case of an accident with vehicles that are both automated and communicating, it is unclear who bears legal responsibility.

## 8.2 IN-VEHICLE ROUTE GUIDANCE

Although it could be subsumed under the text on ADASs, we prefer to discuss in-vehicle route guidance[3] in a separate section, as it is connected more to AI topics than to approaches that arise from the control systems area, typically connected with advanced driver assistance systems.

---

[2]More information on one of these projects (the SARTRE project) can be found at www.sartre-project.eu
[3]Often, these systems are also called "navigation" systems. We prefer the more precise term route guidance. Navigation is a more general term and is used for many aspects involving localization and orientation on many levels of granularity.

As stated in previous chapters, individual routing decisions lead to the traffic state. Therefore, researchers have been investigating the issue of how to influence route choices. Wright [1978] lists five categories of actions for influencing an individual's routing decision with increasing severity: (1) providing route information; (2) giving route advice; (3) pricing roads; (4) imposing access restrictions; and (5) assigning route. At the time when he discussed the potential impacts of route control measures, very primitive systems were tested, in which the road-side infrastructure located at each intersection was tested for sending route information [Boyce, 1988]. Two technological advances were essential: (i) the installation of GNSS (global navigation satellite system) for precise positioning replacing dead reckoning and (ii) the miniaturization of computer memory for fast access to large data sets containing maps stored in RAM instead of on magnetic media. Further developments in display and communication technology, have, since the early 1990s, enabled that commercial in-vehicle route guidance systems, capable of dynamic and adaptive route guidance. This is the case, for example, when a vehicle leaves its planned route or problems are announced via broadcast. Nagaki [2012] gives more details on the evolution of in-vehicle route guidance systems and also shows current development of merging cell phone and entertainment technology with navigation systems.

Figure 8.2 illustrates the major elements necessary for building a route guidance system. The two basic processes are route planning and route checking. Given start and destination positions, and a map containing information about connectivity and costs on edges, a route planning process can determine an optimal route under some user-specific aspect to be minimized. This aspect may be overall travel-time, fuel consumption, etc. From the calculated route, the guidance system produces maneuver information ("turn right," etc.) that are communicated to the user, mostly via a generated speech. While the vehicle is moving, there is continuous monitoring of the position using positioning and map matching. The system checks whether the vehicle is still on the given route, and whether there is some disturbance ahead based on information from the management system (see TMC in Chapter 2) or other external source. If necessary, a new route is computed for the remaining path, and is communicated to the user.

## 8.2.1  BASIC LOCALIZATION

The availability of mobile positioning technology was the essential enabler for modern in-vehicle navigation technology. If the exact position of a vehicle is not known to the routing equipment, no guidance can be provided, no matter on which level of granularity the problem is considered, i.e., neither on the tactical driving level for trajectory following, nor on the strategic level for routing in the road network.

Basically, there are two ways for determining the current position of a vehicle:

1. The position is externally provided, for example by some GNSS, such as GPS (U.S.), Galileo (EU) and corresponding satellite networks. Another source of information on the position and orientation of an vehicle is based on cell phone infrastructure. There are different meth-

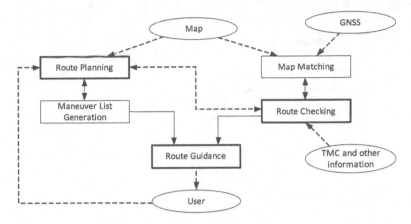

**Figure 8.2:** Schematic view of the elements that lead to a route guidance system and their interaction (adapted from Hofmann-Wellenhof et al., 2003).

ods based on signal strength of the cell to which the phone is currently assigned, and those of the neighboring cells.

2. The position is extrapolated from a last known position using techniques of dead reckoning, in which estimated movement speed and heading are used to estimate the current position.

Practically, both approaches are used. If satellite services are available in sufficient quality, then they are used (currently, they provide position with high precision). If the satellite connection is lost, dead reckoning is used. Positioning is further improved using map matching (see below) for road-bound traffic. Due to the length of literature on navigation, we refer the reader to other sources. Skog and Handel [2012] give a good introduction to different approaches for precise positioning. Nilsson et al. [2012] focus on GNSS. Hofmann-Wellenhof et al. [2003] is an introduction to navigation in general with different methods including mathematical foundations.

## 8.2.2 MAP-MATCHING

Map matching is the processing step that combines basic positioning data with the road network. It determines the position on the link of the network, in which the vehicle is traveling. Hence, it is an essential starting point for (adaptive) route guidance (and of course a number of other location-based services).

As can be seen in Figure 8.3, just matching an isolated measurement to its nearest link is not sufficient; rather, the overall trajectory must be considered. Hofmann-Wellenhof et al. [2003] provide an introduction into basic geometric map-matching. Basically, given a previously matched trajectory and an estimated current position, edge candidates that are near this position

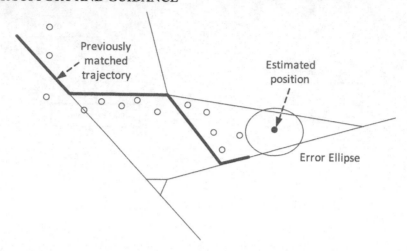

**Figure 8.3:** Illustration of the map matching problem: an estimated position will be linked to the most probably link on a road network. For determining the most probably link, also the previously matched trajectory plays a role.

are evaluated according to how well they fit the previous trajectory. A variety of more advanced algorithms beyond this geometric matching exist. Quddus et al. [2007] provide a general survey. In addition to geometric algorithms, they also list map matching algorithms that use a topological analysis of the network, explicitly involving connectivity. From an AI perspective, interesting algorithms use belief theoretic approaches, like those based on the Dempster-Shafer theory of evidence for handling uncertainty in basic positioning [Nassreddine et al., 2009]. Also, fuzzy logic approaches can be found for improving the performance map-matching. Quddus et al. [2006] suggests a method in which fuzzy rules are used to modify the likelihood of the match. This shows the value of AI based methods, as opposed to those that rely purely on geometry.

### 8.2.3   SHORTEST PATH ALGORITHMS

Based on an accurate estimation of the current position in a network and a destination node, classical shortest path algorithms can be used to calculate the best route between them. We have already briefly mentioned classical shortest path algorithms in Chapter 4. Although A* and Dijkstra are de facto standards in the field, there are mainly two reasons why shortest path algorithms are still an active and highly relevant research topic. First, in in-vehicle route guidance systems, best routes in large scale networks must be calculated extremely fast (within a few seconds) as the user's tolerance for waiting times is low. Second, as travel time is a major ingredient in the heuristics that evaluate individual links, the algorithm must be repeatedly applied to find the currently best (remaining) route to the traveler's destination.

Next we give a brief overview on some of these algorithms for shortest path computation. We assume that standard shortest-path algorithms such as A* or the Dijkstra algorithm are known. A survey on shortest path algorithms adapted for transportation can be found in Fu et al. [2006], also containing information on hierarchical A* (which searches on different levels of aggregation) or bidirectional versions of A* (which starts search at origin and destination nodes at the same time). Buriol et al. [2008] present a comparison of many dynamic shortest path algorithms, and describe a technique that allows the reduction of heap sizes used by several dynamic single-destination shortest-path algorithms.

Both Dijkstra and A* algorithms have a satisfactory performance in static problems. Clearly, the performance of A* depends highly on how well a heuristic estimates costs between origin and destination nodes. When routes must be computed for every driver taking into account the actual cost of each link in the network, the problem is no longer static, given that these links show frequent changes in such costs. Thus, more efficient algorithms must be used. Moreover, in very large networks, there is no time to find the overall optimal route. In this case, anytime algorithms are helpful because a route can be computed, which is the best possible solution given a computing time bound. Anytime algorithms also make sense because it is usually the case that the costs associated with each link will change making expensive computations quickly out-of-date. In such cases it is interesting to compute a partial route that is both fast and inexpensive.

Likhachev [2005] introduced three variants of the A* algorithm: an anytime variant, a dynamic one, and one that is both dynamic and anytime.

- **Anytime Repairing A* (ARA*)** introduces a weight to control the lower bounds of A* and produces a solution with a controlled sub-optimality bound. It finds a sub-optimal solution quickly with a loose weighted bound in the first search. Later, when more time is available, it tightens the bound and reuses previous search efforts until it produces an optimal solution.

- **Lifelong Planning A* (LPA*)** is the dynamic variation. The initial LPA* search is the same used by A*. When there are changes in link costs, LPA* updates them and executes the search again. Subsequent LPA* searches reuse previous valid search efforts to find an updated optimal solution.

- **Anytime Dynamic A*** combines the variable weighted lower bound (ARA* property) and the update of changed costs (LPA* property). The initial solution can be changed after the driver has started his trip, if there were changes in links' costs, or if the initial solution was not the best possible.

As mentioned before, a critical questions concerning the application of A* is the actual shape of the heuristic function. For commercial in-vehicle routing systems, the heuristic is a business secret. Clearly, predicted travel time in a link is a major factor especially when the fastest route is assumed to be the best one.

### 8.2.4   USING ROUTE GUIDANCE

Route guidance systems are prominent applications of advanced traveler information systems (ATIS). Questions about the effect of traffic information on the traffic state have been discussed from a theoretical point of view (see Section 4.3). Broadcasting route information on the current traffic state may be advantageous for a driver, as long as not too many other drivers make the same route choice. Predicting drivers' reactions to ATIS and to routing guidance is an important research area that is becoming more relevant as more drivers use in-vehicle route guidance, more reliable (but possibly identical) information on the current traffic state is available, and more similar route suggestions are given by the different commercial systems.

Yet, a suggested route is relevant for a user only if it fits the current context of the driver and the traffic system. A route may be sub-optimal when the state of the traffic system changes or if the driver has left the route by intentionally or unintentionally taking a wrong turn. The system reacts by quickly calculating an alternative route starting from the current position to the destination. Differences in information on the traffic system as well as an non-transparent heuristic may lead the driver to not trust the system and ignore the route advice given by the navigation system. This issue has been examined in a number of surveys and simulations, such as Chorus et al. [2006] or Klügl and Bazzan [2004a].

## 8.2.5   COOPERATIVE ROUTE GUIDANCE

Currently, in-vehicle route guidance systems generate their information in an isolated way, mostly based on information coming from several external sources such as management centers and location of other users. In the future, perhaps information will come also from other vehicles equipped with V2X communication devices. As discussed in Chapter 4, an efficient decision concerning which route to take depends on others' decisions. In cooperative route guidance systems, calculating and selecting a route is done in a coordinated way, so that each participant travels in a route that is optimal from the system perspective. For instance, Yamashita and Kurumatani [2009] have proposed a cooperative car navigation system with route information sharing. Each vehicle transmits route information (current position, destination, and route to the destination) to a route information server and this estimates future traffic congestion using current congestion information. Estimates are then fed back to each vehicle, which uses them to re-plan its route. Results of the experiments have confirmed that a cooperative car navigation system with the route information sharing generally satisfied individual incentive and social acceptability, as well as had an effect on the improvement of traffic efficiency.

# 8.3 FROM ROUTE GUIDANCE TO TRAVEL RECOMMENDER SYSTEMS

## 8.3.1 EXTENDED ROUTING

A variety of location-based services on mobile phones is available, which extend the scope of traditional routing (in-vehicle devices). Smart phone "apps" are available for finding the nearest pharmacy or bus stop, combined with routing towards the selected point of interest. In the case of public transportation apps, these also include routing information to a destination address. Another good example is the combination of parking and routing: in urban areas, a major problem for travelers using a vehicle concerns where to park after having reached a destination. Bayless and Neelakantan [2012] describe the opportunities of intelligent systems in supporting parking. Whereas systems such as electronic parking payment systems using mobile phones are already available in many cities world-wide (such as the SFPark system provided by the San Francisco Municipal Transportation Agency), systems that lead a driver to the nearest free parking spot are currently available only as prototypic test cases, since they require appropriate sensing equipment to exactly detect free spots, as, for example, the PARKER app by Streetline (`www.streetline.com`). AI technology can be also found in other categories of smart parking systems as, e.g., in systems that control access to a parking lot based on license plate recognition.

An interesting extension of route guidance is the combination with intermediate, spontaneous destination choice, also called "opportunistic routing."[4] Horvitz and Krumm [2012] capture this by providing a road user with information about points of interest that may address a need, while on the way to a primary destination. An example for such a point of interest is a gas station when refueling is needed. Either the user triggers the opportunistic search or sensors in the vehicle may communicate the need. Horvitz et al. [2007] give a general framework for opportunistic planning consisting of (i) means for representing background goals, such as refueling, obtaining groceries, or haircuts; (ii) a method for generating feasible plans for achieving such goals; and (iii) a method for evaluating the utility of alternative plans. For the latter, Horvitz et al. [2007] suggest, for example, the use of Bayesian networks to predict the relevancy of arrival time at the final destination. Bader et al. [2011] show that the acceptance of such suggestions to adapt a route for passing by a gas station is higher, if the particular recommendation comes with sufficient explanation. They argue that pro-activity is necessary due to restricted interaction possibilities and the limited information processing capability of the driver.

## 8.3.2 MOBILE RECOMMENDER SYSTEMS

Whereas classical recommender systems come into play when the user is actively searching for an advice, ubiquitous computing combined with localization technology allows for opportunistic recommendation. Opportunistic routing systems are examples of such mobile recommender sys-

---

[4]The term is more often found in the area of vehicular ad-hoc networks, in which routing of information packages through the wireless network must be opportunistic due to their dynamic structure.

tems: they select an "information item that best suits the needs and preferences of a user in a given situation" [Ricci, 2010]. Items—often products such as books, music, or travel destinations—are ranked based on either matching characteristics of the item with preferences of the user (content-based), or based on what other users with similar interests have chosen before (collaborative filtering). Then, some items are presented to the user. Opportunistic recommendation may be pro-actively given at some point, assuming that the user may need certain items or guidance.

Providing travelers with appropriate recommendations has emerged as one of the major scenarios for recommender systems, even if mainly for leisure travelers and tourists. Recommender systems usually suggest items to users that they did not yet try, but could be interested in. Thus, commuters or experienced travelers are hardly the usual audience for travel recommender systems as they are likely to know their environment very well. Similarly, the opportunistic routing systems introduced in the previous subsection are also particularly interesting to drivers in an unknown network. Nevertheless, they can be also useful in a familiar network to reduce the workload of the driver. For example, given the driver's specific context (location, fuel amount, etc.), it may support the decision on where to best refuel, especially if the recommender system has access to the calendar of the user. Another example for an in-vehicle recommender system whose usefulness is mostly independent on the driver's familiarity with the network is proposed in Baltrunas et al. [2011]. They describe a recommender system for music in a car while driving. In experiments, they have found that user preferences are highly dependent on the context of traffic conditions, as well as on the driver's physical conditions.[5]

In-vehicle recommender systems belong to the category of mobile recommender systems. Ricci [2010] surveys their application potential and particularities.

According to Ricci [2002], some problems of travel recommendation systems are the following. First, travel decisions are complex, so that personal features have to be combined with travel features. Generating destination recommendations alone is feasible. However, for generating really useful recommendations, the travel towards and from the destination has to be integrated into the set of possible options. This is quite costly and depends on the particular application. Further, travel experiences are individual, thus collaborative filtering approaches are not useful. Ricci [2002] compares books with trips. A basic prerequisite for collaborative filtering is that sufficient overlapping of travel experience histories are discovered. Only information on which destination travelers have visited is not enough; rather, information about the context needs to be considered. Ricci suggests more complex approaches for identifying users' needs, such as those based on decision making styles, and multi-stage interactions involving information retrieval about potential options.

---

[5]The experiments were conducted via web-based questionnaires, asking the subject to "Imagine that...," they did not use a driving simulator.

# 8.4   TO KNOW MORE

Data from tracking people's movement using GNSS has been recently acknowledged as an alternative to surveys on route choice or even persons' activities. Also, the mode of transportation can be recognized based on GPS localization and derived traveling speed, for example using a set of heuristic rules as in Gong et al. [2012], or fuzzy rules as in Schüssler and Axhausen [2008]. Based on the identification of stops in the GPS data merged with a network containing possible points of interests, the purpose of a trip can be estimated. Thus, GPS-enabled tools are used in addition to former travel-diary based surveys, making it easy for survey participants to record what they did. Trajectories can be visualized on web-based maps for individual validation, in which the participant annotates the journey with additional information. As an objective measurement, GPS data is expected to be more reliable than self-reporting of survey participants. Stopher et al. [2013] summarize problems of self-reporting in travel surveys and report on a large GPS-based household travel study that avoids any diary self-reporting. Overall, 3,500 participants had to take a GPS device—a personal passive activity logger—with them for three days. After the initial processing of their data, they were asked to validate and fill gaps in the records of derived trips, stops and destinations, travel objectives and modes via Internet, based on a map-based visualization. Thus, this shows that it is both feasible and useful to use mobile GPS technology for large-scale traffic surveys. This revolutionizes the way how data is generated and may lead to new generations of travel demand models allowing for finer resolution and thus better prediction of future travel needs.

Another topic in the area of in-vehicle route guidance systems, where AI techniques are promising, is the intelligent user interface of such systems. It should be as un-intrusive as possible, not distracting the driver. Yet, there is substantial interaction necessary, while potentially driving in stressful, congested traffic. In this case, understanding of natural spoken languages plays a major role.

# CHAPTER 9

# Trends and New Technologies

We are currently experiencing a change in paradigms related to transportation means and traffic engineering, which poses huge challenges to computer science and computer engineering in general, and to AI in particular. The DARPA (Defense Advanced Research Projects Agency) challenge,[1] attracted many research teams around the idea of autonomous driving. This problem poses many challenges not only to vision, perception, and robotics groups, but also to the research on multiagent systems. In fact this seems to be a fertile ground for the design and experimentation of new approaches that address that change in paradigm, as shown by works such as Dresner and Stone [2008] and Vasirani and Ossowski [2011] just to mention two examples.

This book has addressed these and other works that tackle, to different extents, the traditional technologies that underly driving and transportation. These stem mostly from twenty century technologies. In "Reinventing the Automobile," Mitchell et al. [2010] argue that between the two options—(i) "trading off the personal mobility and economic property enabled by automobile transportation to mitigate its negative side effects," and (ii) "take advantage of converging twenty-first century technologies ... to diminish these side effects..."—it is possible to follow the second one. They discuss four ideas to reach a sustainable automobility.

The first is to adopt what the authors call a new automotive DNA. This new DNA opposes: (i) mechanically driven to electrically driven; (ii) powered by internal combustion engine to powered by electric motors; (iii) energized by petroleum to energized by electricity and hydrogen; (iv) mechanically controlled to electronically controlled; and (v) stand-alone operation to intelligent and interconnected. Perhaps it is the last point that relates most closely to computer science and AI. As seen in Chapter 4 (and others) of the present book, stand-alone operation is successful only to a certain extent, as for instance in non-congested situations. In congested regimes, there is a natural coupling between supply and demand. Thus, interconnection in the operation has been already proposed at various levels: from coordinating departure time, to synchronization of traffic signals, to platooning. As discussed in Section 8.1.2, even with today's technology, V2V and GPS allow vehicles to connect and drive together. Indeed, some of the ideas discussed in Mitchell et al. [2010] are not new. What authors propose is their combination in a radically new way.

The second idea—mobility internet—will do for vehicles what the Internet has done for computers: enable vehicles to share enormous amounts of real-time, geo-referenced data, integrating vehicles to the Internet of things, perhaps influencing the travel time optimization methods.

[1]http://archive.darpa.mil/grandchallenge/

The third idea refers to smart, clean energy, while the fourth one is to develop electronically managed, dynamically priced markets for electricity, roads, parking, and vehicles. According to the authors, these markets are underdeveloped today but connectivity can help realize this potential, depending on ubiquitous metering and sensing. Among the various benefits, it would be possible to regulate supply and demand.

At least initially, these four ideas will be (or are already being) implemented in an isolated way. However, the authors make the point that their greatest potential comes when they are pursued together.

## 9.1  INTERCONNECTED AUTOMOBILES

Interconnected vehicles, no matter if electric or mechanically driven, are a trend. For instance, Ford expects to have a wi-fi system on 80% of its cars sold in North America by 2015 (http://www.afr.com/, edition of March 28, 2011), while a surge in worldwide shipments of car wi-fi systems is predicted (to 7.2 million units by 2017, from just 174,000 in 2010). Hence, we focus on characteristics of vehicles that facilitate the interconnection and the realization of the aforementioned agenda.

One possible scenario for interconnected cars has to do with how to reduce congestion. As seem in Chapter /refch:supply, the theory of traffic flow is complex. Traffic streams cross, divide, and combine in complex ways. Only under ideal conditions streams flow smoothly. Either under the near-capacity regime, or when streams are interrupted, human drivers' behaviors become very unstable, because these drivers have to negotiate lane changes, merging, etc. This means that keeping optimal speeds and gaps between vehicles a very hard task for human beings since these often have limited information-processing capacity, are easily distracted, and reach in psychologically complex and sometimes irrational ways [Mitchell et al., 2010]. These authors also mention a study that shows that even small variations and interruptions can propagate shock waves (see Figure 5.2) back through traffic streams for miles, causing traffic jams. This of course means waste of time, space, and energy, not to mention increase in emission of pollutants. Therefore, the investigation about how interconnected vehicles could address these issues in order to optimize and manage traffic flow more efficiently is open. Here, the field of multi-robotics can give a decisive contribution since, for instance, the problems of coordinating swarms of robots and collision avoidance is being investigated there since some time.

According to Mitchell et al., many of the technologies that are needed for connecting vehicles among them and to the infrastructure already exist in commercial forms. There are commercial systems that combine GPS, digital maps, and wireless communication to provide features such as hands-free phones, traffic information, remote diagnostics, etc. In parallel to these connectivity technologies, sensor-based features are also being deployed in vehicles. For example, blind-spot detection can sense objects that may not be visible to the driver; lane-keeping systems help to maintain the lateral position, etc. The participants involved in the aforementioned DARPA challenge have used technologies that combine sensing and GPS, given that communication was not allowed by the rules of the competition. When communication is added to the list of possible technologies, then a vehicle can determine the location and speed of an approaching vehicle, given that current V2V allows each vehicle to broadcast and receive position and speed within a range of a few hundred meters.

For those interested, Mitchell et al. list the advantages and disadvantages of sensing (including image processing) and communication, concluding that both are complementary and there may be a need to combine then in order to have a cost-effective and robust solution. We give a summary here. Sensors tend to add cost and require more hardware. Besides, a vehicle may not be able to sense an object in conditions of poor visibility. On the other hand, communication tends to be cheaper but requires that other vehicles are within the communication range in order to enlarge the perception of a given vehicle.

Regarding sensing of other moving objects in the scenario, there are already transponder-based technologies that allow pedestrians and cyclists to be detected so to eliminate collisions between these participants and other vehicles.

Once the basic task of guaranteeing that vehicles can navigate autonomously without collisions is fulfilled, further steps are to use the connecting infrastructure for other purposes, which range from infrastructure-related to driver/vehicle-centered. In the infrastructure-related category we find smooth out traffic flow, and help traffic signals to coordinate. However, perhaps the most important purposes of connectivity are those that are centered on providing information to the drivers and, ultimately, to the autonomous vehicle itself. In a first stage, assuming that the vehicle is still driven by a human driver, connectivity may help to provide real-time information for drivers. The type of information, again, varies significantly: information about traffic conditions, about toll prices in electronic road pricing, about refueling stations, parking spots, and shopping possibilities are some of the most obvious. Also, in multi-modal transportation, it might be interesting to know which connecting transit possibilities are available where, at which price, etc.

Some of this data can also serve the totally autonomous driven vehicle in a future scenario. For instance, the navigation system can adjust speed and the mix of battery energy and fuel energy in hybrid cars.

In summary, the main message is that electrification provides precise and responsive actuation, while interconnectivity provides situation awareness. By awareness it is meant not only sensor information to avoid immediate crash, but also monthly logs of driving performance, similarly to electricity meters providing logs of energy use, or pressure meters delivering information in health-care applications.

## 9.2    SOME PROJECTS AROUND AUTONOMOUS VEHICLES AND PERSONAL TRANSIT

One of the most known projects of autonomously driven vehicles is the Google car. This project involves developing technology for autonomous cars. According to The New York Times [2010], "during a half-hour drive … a Prius equipped with a variety of sensors and following a route programmed into the GPS navigation system nimbly accelerated in the entrance lane and merged into fast-moving traffic on Highway 101, … The device atop the car produced a detailed map of the environment. The car then drove in city traffic through Mountain View, stopping for lights and stop signs, … The car can be programmed for different driving personalities—from cautious, in which it is more likely to yield to another car, to aggressive, where it is more likely to go first."

We remark that, since then, some U.S. states have passed laws permitting the operation of autonomous vehicles for testing purposes (Nevada, Florida, and California, for example).

However, Google car is hardly the only project on autonomously driven car. Indeed, autonomous devices that resemble a car are well-established in the field of Robotics. Thus, there are many other projects that resemble the Google car. We now discuss some of these, remarking that this list is far from complete.

VisLab's approach to autonomous driving (`http://vislab.it/automotive/`) is based on perception of the surrounding environment in vehicular applications using cameras and other sensors, by means of techniques from artificial vision, image processing, machine learning, neural networks, robotics, and sensor fusion. VisLab has developed a number of vehicle prototypes integrating different functions, from advanced driver assistance systems to automated driving. At the perception level, vehicle, obstacle, pedestrian, and lane detection, as well as traffic signal recognition, and terrain mapping are some examples of the capabilities that have been embedded on the prototypes. At the navigation level, path planning and trajectory planning played basic roles in the development of the vehicles' autonomous capabilities. The current (as 2013) prototype is called BRAiVE (for BRAin-driVE) and was tested recently in the city of Parma, Italy. Before, in 2010, VisLab has conducted the so-called intercontinental autonomous challenge, when another prototype was able to travel 13,000 kms in a 3-month trip from Parma to Shangai (see `www.viac.vislab.it`).

General Motors has a project for a 2-seat urban electric car called EN-V (electric networked-vehicle), which is a jointly developed by Segway Inc. and can be driven normally or operated autonomously. The EN-V was unveiled at the Expo 2010 in Shanghai.

The MIT CityCar is an urban all-electric concept car. This is a light car, which has 60% the size of a Smart Car. It has four independently controlled, electrically powered robotic wheels, which avoids the need for a fixed drive train or axle and enables the car to "fold up" when parked. The top speed is 50 km/h and the all-electric range is 120 km. According to the Financial Times Magazine of May 3, 2013, the MIT Media Lab is working on an autonomous version of the CityCar. The first stage will be a car that parks itself after the occupants have got out and then folds up. The co-director of the City Science research group Kent Larson is quoted saying that "The holy grail is autonomous pick-up and drop-off. You order a car which drives itself to meet you at your chosen point, with the whole fleet controlled by computer."

Therefore, the next frontier to be addressed seems to relate to autonomous vehicles used in car-sharing systems, as well as autonomous personal transit.

As for automated guided personal rapid transit vehicles (also known as podcars or pods), these are already being used today. Pods are used for on-demand, non-stop transportation. In July 2009 the Heathrow Airport launched the ULTRA PRT (personal rapid transit) into operational testing in terminal 5. Since 2010 a similar system operates at Masdar City, UAE.

The next step, in a not-so-distant future, would be to fully implement similar systems in urban environments. Then, the wide use of personal rapid transit vehicles will be associated with a whole agenda of research challenges and technical issues that can now only be anticipated and simulated. For such a simulation, Bazzan et al. [2012] propose an agent-based approach that focus on issues related to the processing of demands for pick-up by autonomous vehicles that serve as a kind of autonomous taxis, as well as on routing them in an efficient way. First, a centralized version is discussed, with a single system manager in charge of processing the whole pick-up demand. Then, extensions deal with decentralized routing, with en-route planning, inter-vehicle communication, and with a market-based approach where the manager of the service runs an auction to determine which customer to serve. All these variants aim at testing alternatives in order to shed light onto questions such as how the podcar service is to be provided, how trips have to be planned, as well as challenges related to the management.

## 9.3 FUTURE OF TRAFFIC MANAGEMENT

### 9.3.1 PARTICIPATORY TRAFFIC MANAGEMENT

Cooperative driver assistance systems based on VLX communications are among the most discussed topics in research on Intelligent Vehicles. Radusch [2013] reports a successful field trial involving 120 vehicles traveling 1.6 million km in central Europe demonstrating the maturity of the technology. V2I technology can lead to a new concept of traffic management as it allows the traffic manager to interact with individual road users. Thus, Radusch shortly sketches the possibility of pro-active mobility management in which drivers and operators cooperate through the overall network of vehicles and infrastructure for achieving an overall optimal functioning of the traffic system. Socially aware drivers cooperate directly with operators who can react to the needs of the all road users immediately. One can even imagine that a vehicle may negotiate with the infrastructure, in order to receive some form of compensation after taking a longer, but less crowded route.

### 9.3.2 AUTONOMIC TRAFFIC MANAGEMENT

An autonomic system can be seen as a system with self-management capabilities. The basic underlying metaphor is taken from the human body, in which a conscious command is accompanied by many low-level self-maintaining processes [Kephart and Chess, 2003]. This general idea is also attractive for traffic network management, in which, often, high-level objectives such as "reducing emissions on a particular road segment" have to be "translated" to adapt signal plans, metering strategies, etc. Currently, this has to be done manually by experiences operators. The recently started EU funded COST-Action "Autonomic Road Traffic Management" (www.cost-arts.org) aims at bringing together researchers from various areas of AI and traffic control for demonstrating the feasibility of such ideas. AI technologies assumed to be enabler for autonomic traffic management are mainly automated planning, learning and adaptive systems, and multiagent systems.

### 9.3.3   CROWD SENSING

Using mobile phones as probes for gathering traffic information forms an approach with huge potential for capturing up-to-date traffic information without further investments in infrastructure (e.g., sensors and cameras). This has been recognized rather early; for a review see [Rose, 2006]. Positioning techniques are key in this domain. There are two ways of positioning: (i) handover from one cell to another (in which accuracy depends on the density of antennae); and (ii) triangulation based on the signal strength from different antennae. Clearly, in addition to that, GNSS-based positioning is possible if the mobile phone possesses a corresponding receiver. Such positioning information can be used not only to determine individual speed and orientation for documenting or predicting travel times, but also to generate traffic state information from on anonymized data from providers of mobile communication or as cellular probes. Starting around 2000, first experiments have been conducted to test the quality of traffic state information generated from mobile phones (summarized in Yim, 2003). Meanwhile, both types of applications are commercially available. For example, the company ESTIMO-TION (www.itistrafficservices.com/), which provides individual services, or Mediamobile (http://www.mediamobile.com/) providing and fusing a wide offer of traffic and weather information-based services.

Information about traffic condition is frequently distributed using new information channels such as Twitter channels. Many municipalities have created their own Twitter channels for communicating traffic conditions, among other types of information. Therefore, it is not surprising that this type of information channels are inspiring researchers to test whether traffic state forecast can be generated from information broadcast via Twitter channels, especially in regions with only little conventional sensing infrastructure. Wanichayapong et al. [2011] show that Twitter messages can, with high accuracy, be classified regarding whether or not they contain information that relates to traffic condition. Whereas they did heavy manual work, Carvalho et al. [2010] showed that a classifier can also be trained using messages from official sources that are then translated into more user-alike messages. Kosala et al. [2012] present a prototypical system for "harvesting" traffic information from Twitter messages, which have exhibited interesting correspondence to the real traffic state, despite the restricted comparison performed. A problem with the current methods is that messages for only a few streets are available. Yet one can expect that with the increasing awareness of the usefulness of such messages, their volume will increase.

# 9.4    TO KNOW MORE

Readers interested on an overview on engineering of electric vehicles are referred to Chapter 1 in Mitchell et al. [2010] as well as Burns et al. [2002].

IEEE has a monthly podcast on ITS topics (IEEE ITS Podcast) at `http://itsp.cicei.com/`.

# Bibliography

Artificial intelligence in transportation – information for application. *Transportation Research Circular E-C113*, 2007. 64

M. Abdoos, N. Mozayani, and A. L.C. Bazzan. Holonic multi-agent system for traffic signals control. *Engineering Applications of Artificial Intelligence*, 26(5–6):1575–1587, 2013. ISSN 0952-1976. DOI: 10.1016/j.engappai.2013.01.007. 70

A. K. Agogino and K. Tumer. A multiagent approach to managing air traffic flow. *Autonomous Agents and Multi-Agent Systems*, 24(1):1–25, 2012. DOI: 10.1007/s10458-010-9142-5. xiv

T. Arentze and H. Timmermans. Albatross: overview of the model, application and experiences. In *Innovation in Travel Modeling 2008 Conference*, 2008. 29

T. Arentze and H. Timmermans. A learning-based transportation oriented simulation system. *Transportation Research Part B: Methodological*, 38(7):613 – 633, 2004. DOI: 10.1016/j.trb.2002.10.001. 29

R. Arnott and K. Small. The economics of traffic congestion. *American Scientist*, 82:446–455, 1994. 22

R. Arnott, A. de Palma, and R. Lindsey. Does providing information to drivers reduce traffic congestion? *Transportation Research A*, 25:309–318, 1991. DOI: 10.1016/0191-2607(91)90146-H. 40, 43

Tsz-Chiu Au, N. Shahidi, and P. Stone. Enforcing liveness in autonomous traffic management. In W. Burgard and D. Roth, Editors, *Proceedings of the Twenty-Fifth Conference on Artificial Intelligence*, pages 1317–1322. AAAI Press, August 2011. 73

K. Axhausen and T. Gärling. Activity-based approaches to travel analysis: conceptual frameworks, models and research problems. *Transportation Reviews*, 12:323–341, 1992. DOI: 10.1080/01441649208716826. 30

R. Bader, O. Siegmund, and W. Woerndl. A study on user acceptance of proactive in-vehicle recommender systems. In *Proceedings of the 3rd International Conference on Automotive User Interfaces and Interactive Vehicular Applications*, pages 47–54. ACM, 2011. DOI: 10.1145/2381416.2381424. 87

G. Balan and S. Luke. History-based traffic control. In H. Nakashima, M. P. Wellman, G. Weiss, and P. Stone, Editors, *Proceedings of the Fifth International Joint Conference on Autonomous Agents and Multiagent Systems*, pages 616–621, New York, NY, USA, 2006. ACM Press. ISBN 1-59593-303-4. DOI: 10.1145/1160633. 73

M. Balmer, M. Rieser, K. Meister, D. Charypar, N. Lefebre, and K. Nagel. MATSim-T: Architecture and simulation times. In A. L. Bazzan and F. Klügl, Editors, *Multi-Agent Systems for Traffic and Transportation Engineering*, pages 57–78. IGI Global, Hershey, US, 2009. ISBN 978-1-60566-226-8. DOI: 10.4018/978-1-60566-226-8. 29, 59

M. Balmer, N. Cetin, K. Nagel, and B. Raney. Towards truly agent-based traffic and mobility simulations. In N. R. Jennings, C. Sierra, L. Sonenberg, and M. Tambe, Editors, *Proceedings of the 3rd International Joint Conference on Autonomous Agents and Multi Agent Systems, AAMAS*, volume 1, pages 60–67, New York, USA, July 2004. New York, IEEE Computer Society. 60, 61, 63

L. Baltrunas, M. Kaminskas, B. Ludwig, O. Moling, F. Ricci, A. Aydin, K. Lüke, and R. Schwaiger. Incarmusic: Context-aware music recommendation in a car. *12th International Conference on Electronic Commerce and Web Technologies (EC-Web)*, pages 89–100, 2011. DOI: 10.1007/978-3-642-23014-1_8. 88

J. Barceló, editor. *Fundamentals of Traffic Simulation*. Springer, 2010. DOI: 10.1007/978-1-4419-6142-6. 24, 53, 56

S. H. Bayless and R. Neelakantan. Smart parking and the connected consumer – opportunities for facility operators and municipalities. Technical report, The Intelligent Transportation Society of America (ITS America), 2012. 87

A. L. C. Bazzan. A distributed approach for coordination of traffic signal agents. *Autonomous Agents and Multiagent Systems*, 10(1):131–164, March 2005. DOI: 10.1007/s10458-004-6975-9. 67

A. L. C. Bazzan. Opportunities for multiagent systems and multiagent reinforcement learning in traffic control. *Autonomous Agents and Multiagent Systems*, 18(3):342–375, June 2009. DOI: 10.1007/s10458-008-9062-9. 75

A. L. C. Bazzan and F. Klügl. Re-routing agents in an abstract traffic scenario. In G. Zaverucha and A. Loureiro da Costa, Editors, *Advances in artificial intelligence*, number 5249 in Lecture Notes in Artificial Intelligence, pages 63–72, Berlin, 2008. Springer-Verlag. 71

A. L. C. Bazzan and F. Klügl. A review on agent-based technology for traffic and transportation. *The Knowledge Engineering Review*, FirstView:1–29, 4 2013. ISSN 1469-8005. DOI: 10.1017/S0269888913000118. 60, 75

A. L. C. Bazzan, J. Wahle, and F. Klügl. Agents in traffic modelling - from reactive to social behavior. In *Advances in Artificial Intelligence*, number 1701 in Lecture Notes in Artificial Intelligence, pages 303–306, Berlin/Heidelberg, 1999. Springer. Extended version appeared in Proc. of the U.K. Special Interest Group on Multi-Agent Systems (UKMAS), Bristol, UK. DOI: 10.1007/3-540-48238-5_28. 60, 61, 63

A. L. C. Bazzan, D. de Oliveira, F. Klügl, and K. Nagel. Adapt or not to adapt – consequences of adapting driver and traffic light agents. In K. Tuyls, A. Nowe, Z. Guessoum, and D. Kudenko, Editors, *Adaptive Agents and Multi-Agent Systems III*, volume 4865 of *Lecture Notes in Artificial Intelligence*, pages 1–14. Springer-Verlag, 2008. ISBN 978-3-540-77947-6. DOI: 10.1007/978-3-540-77949-0. 71

A. L. C. Bazzan, D. de Oliveira, and B. C. da Silva. Learning in groups of traffic signals. *Eng. Applications of Art. Intelligence*, 23:560–568, 2010. DOI: 10.1016/j.engappai.2009.11.009. 70

A. L. C. Bazzan, M. B. Amarante, and F. Beschoren da Costa. Management of demand and routing in autonomous personal transportation. *Journal of Intelligent Transportation Systems*, 16(1):1–11, 2012. DOI: 10.1080/15472450.2012.639635. 75, 95

M. Ben-Akiva and S. Lerman. *Discrete Choice Analysis – Theory and Application to Travel Demand*. MIT Press, 1985. 30

M. Ben-Akiva and M. Bierlaire. Discrete choice methods and their applications to short term travel decisions. In *Handbook of transportation science*, pages 5–33. Springer US, 1999. DOI: 10.1007/978-1-4615-5203-1_2. 30

M. Ben-Akiva, H. N. Koutsopoulos, C. Antoniou, and R. Balakrishna. *Traffic Simulation with DynaMIT*, chapter 10. Springer, 2010. 59

C. R. Bhat and F. S. Koppelman. A retrospective and prospective survey of time-use research. *Transportation*, 26:119–139, 1999. DOI: 10.1023/A:1005196331393. 30

M. Bielli, G. Ambrosino, and M. Boero, Editors. *Artificial Intelligence Application in Traffic Engineering*. VSP, Zeist, NL, 1994. 64

J. A. Bonneson, S. R. Sunkari, M. P. Pratt, and Texas Transportation Institute and Texas Department of Transportation and United States Federal Highway Administration. Traffic signal operations handbook. Technical Report FHWA/TX-09/0-5629-P1, Texas Transportation Institute and Texas Department of Transportation and United States Federal Highway Administration, 2009. http://d2dtl5nnlpfr0r.cloudfront.net/tti.tamu.edu/documents/0-5629-P1.pdf. 17, 22

D. E. Boyce. Route guidance systems for improving urban travel and location choices. *Transportation Research A*, 22(4):275–281, 1988. DOI: 10.1016/0191-2607(88)90005-2. 82

M. Brackstone and M. McDonald. Car–following: a historical review. *Transportation Res. F*, 2: 181–196, 2000. DOI: 10.1016/S1369-8478(00)00005-X. 79

D. Braess. Über ein Paradoxon aus der Verkehrsplanung. *Unternehmensforschung*, 12:258, 1968. DOI: 10.1007/BF01918335. 41, 42

W. Burghout. *Hybrid Micro-Mesoscopic traffic simulation*. Ph.D. thesis, Royal Institute of Technology, Stockholm, Sweden, 2004. 59

W. Burghout, H. Koutsopoulos, and I. Andreasson. Hybrid mesoscopic-microscopic traffic simulation. *Transportation Research Records*, 1934:218–225, 2005. DOI: 10.3141/1934-23. 59

L. S. Buriol, M. G. C. Resende, and M. Thorup. Speeding up dynamic shortest-path algorithms. *INFORMS Journal on Computing*, 20(2):191–204, 2008. DOI: 10.1287/ijoc.1070.0231. 85

L. S. Buriol, M. Ritt, F. Rodrigues, and G. Schäfer. On the smoothed price of anarchy of the traffic assignment problem. In *ATMOS*, pages 122–133, 2011. DOI: 10.4230/OASIcs.ATMOS.2011.122. 39

L. D. Burns, J. B. McCormick, and C. E. Borroni-Bird. Vehicle of change. *Scientific American*, 287(4):65–73, October 2002. DOI: 10.1038/scientificamerican1002-64. 97

E. Camponogara and W. Kraus Jr. Distributed learning agents in urban traffic control. In F. Moura-Pires and S. Abreu, Editors, *EPIA*, pages 324–335, 2003. 68

S. Carvalho, L. Sarmento, and R. J. F. Rossetti. Real-time sensing of traffic information in Twitter messages. In *Proceedings of the IEEE ITSC 2010 Workshop on Artificial Transportation Systems and Simulation (ATSS-2010)*, 2010. 96

E. Cascetta. *Transportation Systems Analysis – Models and Applications*, volume 29 of *Springer Optimization and Its Applications*. Springer, 2009. 27, 33, 38, 42

E. Chan, P. Gilhead, P. Jelinek, P. Krejci, and T. Robinson. Cooperative control of SARTRE automated platoon vehicles. In *Proceedings of the 19th ITS World Congress*, 2012. 81

E. C. P. Chang, J. C. K. Lei, and C. J. Messer. Arterial signal timing optimization using PASSER-II. Technical Report 467, Texas Transportation Institute, 1988. 21

B. Chen and H. H. Cheng. A review of the applications of agent technology in traffic and transportation systems. *IEEE Transactions in Intelligent Transportation Systems*, 11(2):485–497, 2010. DOI: 10.1109/TITS.2010.2048313. 75

C. Chen, F. Zhu, and Y. Ai. A survey of urban traffic signal control for agent recommendation system. In *15th International IEEE Conference on Intelligent Transportation Systems (ITSC)*, pages 327–333, 2012. DOI: 10.1109/ITSC.2012.6338604. 75

Y.-C. Chiu, J. Bottom, M. Mahut, A. Paz, R. Balakrishna, T. Waller, and J. Hicks. Dynamic traffic assignment: A primer. Transportation Research Circular E-C153, June 2011. 37, 43

T. Chmura and T. Pitz. An extended reinforcement algorithm for estimation of human behavior in congestion games. *Journal of Artificial Societies and Social Simulation*, 10(2), 2007. 41

C. G. Chorus, E. J. Molin, and B. van Wee. Use and effects of advanced traveller information services (atis): A review of the literature. *Transport Reviews*, 26(2):127–149, 2006. DOI: 10.1080/01441640500333677. 86

J. Y. Chow, C. H. Yang, and A. C. Regan. State-of-the art of freight forecast modeling: lessons learned and the road ahead. *Transportation*, 37:1011–1030, 2010. DOI: 10.1007/s11116-010-9281-1. 32

M. C. Choy, D. Srinivasan, and R. L. Cheu. Cooperative, hybrid agent architecture for real-time traffic signal control. *IEEE Transaction on Systems, Man and Cybernetics- Part 1: Systems and Humans*, 33(5):597–607, 2003. DOI: 10.1109/TSMCA.2003.817394. 70

T. G. Crainic and G. Laport. Planning models for freight transportation. *European Journal of Operational Research*, 97:409–438, 1997. DOI: 10.1016/S0377-2217(96)00298-6. 32

C. F. Daganzo and Y. Sheffi. On stochastic models of traffic assignment. *Transportation Science*, 11(3):253–274, 1977. DOI: 10.1287/trsc.11.3.253. 37, 42

P. Davidsson, L. Henesey, L. Ramstedt, J. Törnquist, and F. Wernstedt. An analysis of agent-based approaches to transport logistics. *Transportation Research C*, 13:255–271, 2005. DOI: 10.1016/j.trc.2005.07.002. xiv, 32

D. de Oliveira, P. R. Ferreira, Jr., and A. L. C. Bazzan. A swarm based approach for task allocation in dynamic agents organizations. In *Proceedings of the 3rd International Joint Conference on Autonomous Agents and Multi Agent Systems, AAMAS*, volume 3, pages 1252–1253, New York, USA, July 2004. IEEE Computer Society. DOI: 10.1109/AAMAS.2004.33. 67

L. Barcelos de Oliveira and E. Camponogara. Multi-agent model predictive control of signaling split in urban traffic networks. *Transportation Research Part C: Emerging Technologies*, 18(1): 120–139, 2010. ISSN 0968-090X. DOI: 10.1016/j.trc.2009.04.022. 21, 66

J. J. de Vries, P. Nijkamp, and P. Rietveld. Alonoso's theory of movements: Developments in spatial interaction modeling. *Journal of Geographical Systems*, 3:233–256, 2001. DOI: 10.1007/PL00011478. 27

A. dePalma, R. Lindsey, and N. Picard. Risk aversion, the value of information, and traffic equilibrium. *Transportation Science*, 46(1):1–26, 2012. DOI: 10.1287/trsc.1110.0357. 43

C. Desjardins, J. Laumônier, and B. Chaib-draa. Learning agents for collaborative driving. In A. L. C. Bazzan and F. Klügl, Editors, *Multi-Agent Systems for Traffic and Transportation*, pages 240–260. IGI Global, Hershey, PA, 2009. ISBN 978-1-60566-226-8. DOI: 10.4018/978-1-60566-226-8. 81

M. Di Taranto. UTOPIA. In *Proc. of the IFAC-IFIP-IFORS Conference on Control, Computers, Communication in Transportation*, pages 245–252, Paris, 1989. International Federation of Automatic Control. 21

H. Dia. An agent-based approach to modeling driver route choice behaviour under the influence of real-time information. *Transportation Research C*, 10(5-6):331–349, 2002. DOI: 10.1016/S0968-090X(02)00025-6. 63

C. Diakaki, M. Papageorgiou, and K. Aboudolas. A multivariable regulator approach to traffic-responsive network-wide signal control. *Control Engineering Practice*, 10(2):183–195, February 2002. DOI: 10.1016/S0967-0661(01)00121-6. 21

R. B. Dial. A probabilistic multi-path traffic assignment algorithm. *Transportation Research*, 5: 83–111, 1971. DOI: 10.1016/0041-1647(71)90012-8. 35

R. B. Dial. Minimal-revenue congestion pricing part I: A fast algorithm for the single-origin case. *Transportation Research Part B: Methodological*, 33(3):189–202, 1999. DOI: 10.1016/S0191-2615(98)00026-5. 22

R. B. Dial. Minimal-revenue congestion pricing part II: An efficient algorithm for the general case. *Transportation Research Part B: Methodological*, 34(8):645–665, 2000. DOI: 10.1016/S0191-2615(99)00046-6. 22

E. W. Dijkstra. A note on two problems in connection with graphs. *Numerische Mathematik*, 1: 269–271, 1959. DOI: 10.1007/BF01386390. 34

A. Doniec, R. Mandiau, S. Piechowiak, and S. Espié. A behavioral multi-agent model for road traffic simulation. *Engineering Applications of Artificial Intelligence*, 21(8):1443–1454, 2008. DOI: 10.1016/j.engappai.2008.04.002. 62

K. Dresner and P. Stone. Multiagent traffic management: A reservation-based intersection control mechanism. In N. R. Jennings, C. Sierra, L. Sonenberg, and M. Tambe, Editors, *Proc. of the International Joint Conference on Autonomous Agents and Multiagent Systems*, pages 530–537, New York, USA, July 2004, IEEE Computer Society. 72

K. Dresner and P. Stone. Multiagent traffic management: Opportunities for multiagent learning. In K. Tuyls, P. J. Hoen, K. Verbeeck, and S. Sen, Editors, *LAMAS 2005*, number 3898 in Lecture Notes in Artificial Intelligence, pages 129–138. Springer Verlag, Berlin, 2006. 73

K. Dresner and P. Stone. A multiagent approach to autonomous intersection management. *Journal of Artificial Intelligence Research*, 31:591–656, March 2008. DOI: 10.1613/jair.2502. 73, 91

A. Eskandarian, editor. *Handbook of Intelligent Vehicles*. Springer, London, 2012. DOI: 10.1007/978-0-85729-085-4. 79

S. Espie and J. M Auberlet. ARCHISIM: a behavioural multi-actor traffic simulation model for the study of a traffic system including its aspects. *International Journal of ITS Research*, 5(1): 7–16, 2007. 61, 62

J. Esser and M. Schreckenberg. Microscopic simulation of urban traffic based on cellular automata. *International Journal of Modern Physics C*, 8(5):1025, 1997. DOI: 10.1142/S0129183197000904. 57

M. Fellendorf and P. Vortisch. Microscopic traffic flow simulator VISSIM. In J. Barcelo, editor, *Fundamentals of Traffic Simulation*, pages 63–94. Springer, New York, 2010. DOI: 10.1007/978-1-4419-6142-6. 56

J. France and A. A. Ghorbani. A multiagent system for optimizing urban traffic. In *Proceedings of the IEEE/WIC International Conference on Intelligent Agent Technology*, pages 411–414, Washington, DC, USA, 2003. IEEE Computer Society. ISBN 0-7695-1931-8. DOI: 10.1109/IAT.2003.1241110. 67

M. Frank and P. Wolfe. An algorithm for quadratic programming. *Naval Research Logistics Quarterly*, 3(1–2):95–110, 1956. DOI: 10.1002/nav.3800030109. 36, 39

L. Fu, D. Sun, and L. R. Rilett. Heuristic shortest path algorithms for transportation applications: State of the art. *Computers & Operations Research*, 33(11):3324–3343, 2006. DOI: 10.1016/j.cor.2005.03.027. 85

S. Md. Galib and I. Moser. Road traffic optimisation using an evolutionary game. In *Proceedings of the 13th annual conference companion on Genetic and evolutionary computation*, GECCO '11, pages 519–526, New York, NY, USA, 2011. ACM. ISBN 978-1-4503-0690-4. DOI: 10.1145/2001858.2002043. 41

N. H. Gartner. OPAC - a demand-responsive strategy for traffic signal control. *Transportation Research Record*, 906:75–81, 1983. 21

N. H. Gartner, C.J. Messer, and A.K. Rathi. *Traffic flow theory: A state-of-the-art report*. Federal Highway Administration, 2001. http://www.fhwa.dot.gov/publications/research/operations/tft/. 2, 12, 22

C. Gawron. *Simulation-based traffic assignment*. Ph.D. thesis, University of Cologne, Cologne, Germany, 1998a. www.zaik.uni-koeln.de/AFS/publications/theses.html. 38, 39, 59

## 106 BIBLIOGRAPHY

C. Gawron. An iterative algorithm to determine the dynamic user equilibrium in a traffic simulation model. *International Journal of Modern Physics C*, 9(3):393–407, 1998b. DOI: 10.1142/S0129183198000303 . 38

G. Gentile. Local user cost equilibrium: a bush-based algorithm for traffic assignment. *Transportmetrica*, iFirst:1–40, 2012. DOI: 10.1080/18128602.2012.691911. 36

D. L. Gerlough and M. J. Huber. *Traffic flow theory: a monograph*. Transportation research board. National research council, Washington, DC, 1975. ISBN 0-309-02459-5. 22

C. Gershenson. *Design and Control of Self-organizing Systems*. Ph.D. thesis, Vrije Universiteit Brussel, Brussels, Belgium, May 2007. http://cogprints.org/5442/. 66, 67

P. G. Gipps. A model for the structure of lane-changing decisions. *Transportation Research Part B: Methodological*, 20(5):403–414, 1986. DOI: 10.1016/0191-2615(86)90012-3. 64

P. G. Gipps. A behavioural car-following model for computer simulation. *Transportation Research Part B: Methodological*, 15(2):105–111, 1981. DOI: 10.1016/0191-2615(81)90037-0. 56, 64

H. Gong, C. Chen, E. Bialostozky, and C. T. Lawson. A GPS/GIS method for travel mode detection in new york city. *Computers, Environment and Urban Systems*, 36(2):131–139, 2012. DOI: 10.1016/j.compenvurbsys.2011.05.003. 89

R. L. Gordon and W. Tighe. Traffic control systems handbook. Technical Report FHWA-HOP-06-006, Office of Transportation Management, Federal Highway Administration, Washington, DC, 2005. ops.fhwa.dot.gov/publications/fhwahop06006/. 367 pages. 22, 75

B. Greenshields. A study of traffic capacity. In *Proc. of Highway Res. Board*, pages 448–477, vol. 14, 1935. 11

IHRA ITS Working Group. Design principles for advanced driver assistance systems: Keeping drivers in-the-loop, December 2010. 79

C. Guestrin, M. G. Lagoudakis, and R. Parr. Coordinated reinforcement learning. In C. Sammut and A. G. Hoffmann, Editors, *Proceedings of the Nineteenth International Conference on Machine Learning (ICML)*, pages 227–234, San Francisco, CA, USA, 2002. Morgan Kaufmann. ISBN 1-55860-873-7. 70

W. Hajek, I. Gaponova, K. H. Fleischer, and J. Krems. Workload-adaptive cruise control – a new generation of advanced driver assistance systems. *Transportation Research Part F: Traffic Psychology and Behaviour*, 20:108–120, 2013. DOI: 10.1016/j.trf.2013.06.001. 79

M. Haklay. How good is volunteered geographical information? a comparative study of OpenStreetMap and Ordnance Survey datasets. *Environment and Planning B: Planning and Design*, 37(4):682–703, 2010. DOI: 10.1068/b35097. 51

Martin L. Hazelton. Some remarks on stochastic user equilibrium. *Transportation Research Part B: Methodological*, 32(2):101–108, February 1998. DOI: 10.1016/S0191-2615(97)00015-5. 42

D. Helbing, S. Lämmer, and P. Lebacque. Self-organized control of irregular or perturbed network traffic. In C. Deissenberg and R.F. Hartl, Editors, *Optimal Control and Dynamic Games*, page 239. Springer, 2005. DOI: 10.1007/b136166. 67

D. Helbing. Traffic and related self-driven many-particle systems. *Rev. Mod. Phys.*, 73:1067–1141, December 2001. DOI: 10.1103/RevModPhys.73.1067. 22

J. Henry, J. L. Farges, and J. Tuffal. The PRODYN real time traffic algorithm. In R. Isermann, editor, *Proceedings of the Int. Fed. of Aut. Control (IFAC) Conf.*, pages 307–312, Baden-Baden, 1983. IFAC. 21

P. Y. Herzberg. Beyond "accident-proneness": Using five-factor model prototypes to predict driving behavior. *Journal of Research in Personality*, 43(6):1096 – 1100, 2009. DOI: 10.1016/j.jrp.2009.08.008. 62

B. Hofmann-Wellenhof, K. Legat, and M. Wieser. *Navigation - Principles of Positioning and Guidance*. Springer, Wien, 2003. 83

J. Holmgren, P. Davidsson, J. A. Persson, and L. Ramstedt. Tapas: A multi-agent-based model for simulation of transport chains. *Simulation Modelling Practice and Theory*, 23:1–18, 2012. DOI: 10.1016/j.simpat.2011.12.011. 32

S. P. Hoogendoorn and P. H. L. Bovy. State-of-the-art of vehicular traffic flow modelling. *Journal of Systems and Control Engineering*, 215(4):283–303, 2001. DOI: 10.1177/095965180121500402. 60

E. Horvitz, P. Koch, and M. Subramani. Mobile opportunistic planning: Methods and models. In C. Conati, K. McCoy, and G. Paliouras, Editors, *User Modelling 2007*, volume 4511 of *Lecture Notes in Computer Science*, pages 228–237. Springer Berlin Heidelberg, 2007. DOI: 10.1007/978-3-540-73078-1. 87

E. Horvitz and J. Krumm. Some help on the way: Opportunistic routing under uncertainty. *UbiComp' 12*, pages 1–10, July 2012. DOI: 10.1145/2370216.2370273. 87

P. B. Hunt, D. I. Robertson, R. D. Bretherton, and R. I. Winton. SCOOT - a traffic responsive method of coordinating signals. TRRL Lab. Report 1014, Transport and Road Research Laboratory, Berkshire, 1981. 21

J. Illenberger. *Social Networks and Cooperative Travel Behaviour*. Ph.D. thesis, TU Berlin, Faculty V - Verkehrs- und Maschinensysteme, Germany, 2012. 64

J. W. Joubert, P. J. Fourie, and K. W. Axhausen. Large-scale agent-based combined traffic simulation of private cars and commercial vehicles. *Transportation Research Record*, (2168):24–32, 2010. DOI: 10.3141/2168-04. 32

R. Junges and A. L. C. Bazzan. Evaluating the performance of DCOP algorithms in a real world, dynamic problem. In L. Padgham, D. Parkes, J. Müller, and S. Parsons, Editors, *Proc. of the 7th Int. Joint Conf. on Aut. Agents and Multiagent Systems*, pages 599–606. IFAAMAS, May 2008. 66

M. S. Kakkasageri and S. S. Manvi. Information management in vehicular ad hoc networks: A review. *Journal of Network and Computer Applications*, 2013. DOI: 10.1016/j.jnca.2013.05.015. 80

J. O. Kephart and D. M. Chess. The vision of autonomic computing. *IEEE Computer*, 36(1): 41–50, 2003. DOI: 10.1109/MC.2003.1160055. 95

B. S. Kerner, S. L. Klenov, and D. E. Wolf. Cellular automata approach to three-phase traffic theory. *Journal of Physics A: Mathematical and General*, 35:9971–20013, 2002. DOI: 10.1088/0305-4470/35/47/303. 57

B. S. Kerner. *Introduction to Modern Traffic Flow Theory and Control*. Springer London, Limited, 2009. ISBN 9783642026058. DOI: 10.1007/978-3-642-02605-8. 12, 22, 48

A. Kesting, M. Treiber, and D. Helbing. Agents for traffic simulation. In A. M. Uhrmacher and D. Weyns, Editors, *Multi-Agent Systems: Simulation and Applications*, pages 325–356. CRC Press, Boca Raton, 2010. 56, 60

F. Klügl and A. L. C. Bazzan. Simulation studies on adaptative route decision and the influence of information on commuter scenarios. *Journal of Intelligent Transportation Systems: Technology, Planning, and Operations*, 8(4):223–232, October–December 2004a. DOI: 10.1080/15472450490523874. 41, 86

F. Klügl and A. L. C. Bazzan. Route decision behaviour in a commuting scenario. *Journal of Artificial Societies and Social Simulation*, 7(1), 2004b. jasss.soc.surrey.ac.uk/7/1/1. html. 41

F. Klügl and A. L. C. Bazzan. Agent-based modeling and simulation. *AI Magazine*, 33(3):29–40, 2012. http://www.aaai.org/ojs/index.php/aimagazine/article/view/2425. 53

F. Klügl and G. Rindsfüser. Agent-based route (and mode) choice simulation in real-world networks. In *Web Intelligence and Intelligent Agent Technology (WI-IAT), 2011 IEEE/WIC/ACM International Conference on*, pages 22–29, Lyon, France, 2011. DOI: 10.1109/WI-IAT.2011.246. 63

W. Knospe, L. Santen, A. Schadschneider, and M. Schreckenberg. Towards a realistic microscopic description of highway traffic. *J. Phys. A*, 33(48):L477, 2000. DOI: 10.1088/0305-4470/33/48/103. 61

R. Kosala, E. Adi, and Steven. Harvesting real time traffic information from twitter. *Procedia Engineering*, 50:1–11, 2012. DOI: 10.1016/j.proeng.2012.10.001. 96

I. Kosonen. Multi-agent fuzzy signal control based on real-time simulation. *Transportation Research C*, 11(5):389–403, 2003. DOI: 10.1016/S0968-090X(03)00032-9. 66

A. Kotsialos, M. Papageorgiou, C. Diakaki, Y. Pavlis, and F. Middelham. Traffic flow modeling of large scale motorway networks using the macroscopic modeling tool metanet. *IEEE Transactions on Intelligent Transportation Systems*, 3(4):282–292, 2002. DOI: 10.1109/TITS.2002.806804. 58

E. Koutsoupias and C. Papadimitriou. Worst-case equilibria. In *Proceedings of the 16th annual conference on Theoretical aspects of computer science (STACS)*, pages 404–413, Berlin, Heidelberg, 1999. Springer-Verlag. ISBN 3-540-65691-X. DOI: 10.1016/j.cosrev.2009.04.003. 40, 42

J. Koźlak, G. Dobrowolski, M. Kisiel-Dorohinicki, and E. Nawarecki. Anti-crisis management of city traffic using agent-based approach. *Journal of Universal Computer Science*, 14(14):2359–2380, 2008. DOI: 10.3217/jucs-014-14-2359. 66

D. Krajzewicz, J. Erdmann, M. Behrisch, and L. Bieker. Recent development and applications of SUMO - Simulation of Urban MObility. *International Journal On Advances in Systems and Measurements*, 5(3&4):128–138, 2012. 56

L. Kuyer, S. Whiteson, B. Bakker, and N. A. Vlassis. Multiagent reinforcement learning for urban traffic control using coordination graphs. In *ECML/PKDD*, pages 656–671, Antwerp, Belgium, 2008. DOI: 10.1007/978-3-540-87479-9_61. 69

B. Lenntorp. Time geography - at the end of its beginning. *GeoJournal*, 48:155–158, 1999. DOI: 10.1023/A:1007067322523. 28

W. Leutzbach. *Introduction to the theory of traffic flow*. Springer, 1988. ISBN 9783540171133. DOI: 10.1007/978-3-642-61353-1. 12, 22, 48

D. Levinson. The value of advanced traveler information systems for route choice. *Transportation Research Part C: Emerging Technologies*, 11(1):75–87, 2003. DOI: 10.1016/S0968-090X(02)00023-2. 43

G. Liedtke. Principles of micro-behavior commodity transport modeling. *Transportation Research Part E: Logistics and Transportation Review*, 45(5):795–809, 2009. DOI: 10.1016/j.tre.2008.07.002. 32

M. Likhachev. *Search-based Planning for Large Dynamic Environments*. Ph.D. thesis, Carnegie Mellon University, 2005. 85

C. Lochert, J. Rybicki, B. Scheuermann, and M. Mauve. Scalable data dissemination for inter-vehicle-communication: Aggregation versus peer-to-peer. *Oldenbourg IT – Information Technology*, 50(4):237–242, August 2008. DOI: 10.1524/itit.2008.0490. 81

P. Lowrie. The Sydney coordinate adaptive traffic system - principles, methodology, algorithms. In *Proceedings of the International Conference on Road Traffic Signalling*, Sydney, Australia, 1982. 21

Y. Luo and L. Bölöni. Modeling the conscious behavior of drivers for multi-lane highway driving. *Proc. of the 7th Workshop on Agents in Traffic and Transportation at AAMAS 2012*, 2012. 64

C. Maag, C. Mark, and H.-P. Krüger. Development of a cognitive-emotional model for driver behavior. In *Agent and multi-agent systems: Technologies and applications, Springer*, pages 242–251, 2010. DOI: 10.1007/978-3-642-13541-5_25. 63

H. Mahmassani and R. Herman. Dynamic user equilibrium departure time and route choice on idealized traffic arterials. *Transportation Science*, 18(4):362–384, 1984. DOI: 10.1287/trsc.18.4.362. 38

M. Mahut and M. Florian. Traffic simulation with Dynameq. In J. Barcelo, editor, *Fundamentals of Traffic Simulation*, pages 323–362. Springer, New York, 2010. DOI: 10.1007/978-1-4419-6142-6. 59

R. Mandiau, A. Champion, J.-M. Auberlet, S. Espié, and C. Kolski. Behaviour based on decision matrices for a coordination between agents in a urban traffic simulation. *Appl. Intell.*, 28(2):121–138, 2008. DOI: 10.1007/s10489-007-0045-3. 72

H. J. Miller. Modelling accessibility using space-time prism concepts within geographical information systems. *International journal of geographical information systems*, 5(3):287–301, 1991. DOI: 10.1080/02693799108927856. 28

W. J. Mitchell, C. E. Borroni-Bird, and L. D. Burns. *Reinventing the Automobile*. MIT Press, Cambridge, MA, 2010. 91, 92, 93, 97

J. T. Morgan and J. D. C. Little. Synchronizing traffic signals for maximal bandwidth. *Operations Research*, 12:897–912, 1964. DOI: 10.1287/opre.12.6.896. 19

E. K. Morlok. *Introduction to transportation engineering and planning*. McGraw-Hill, 1978. ISBN 9780070431324. xiv, 2, 9

K. Nagaki. Evolution of in-car navigation systems. In A. Eskandarian, editor, *Handbook of Intelligent Vehicles*, pages 464–487. Springer, London, 2012. DOI: 10.1007/978-0-85729-085-4. 82

K. Nagel and M. Schreckenberg. A cellular automaton model for freeway traffic. *Journal de Physique I*, 2:2221, 1992. DOI: 10.1051/jp1:1992277. 57

K. Nagel. Personal Communication, 2007. 64

G. Nassreddine, F. Abdallah, and T. Denoeux. Map matching algorithm using interval analysis and dempster-shafer theory. In *Intelligent Vehicles Symposium, 2009 IEEE*, pages 494–499, 2009. DOI: 10.1109/IVS.2009.5164328. 84

The New York Times. Google cars drive themselves, in traffic., October 9 2010. http://www.nytimes.com/2010/10/10/science/10google.html?_r=2&. accessed August, 2013. 93

J.-O. Nilsson, D. Zachariah, and I. Skog. Global navigation satellite systems: An enabler for in-vehicle navigation. In A. Eskandarian, editor, *Handbook of Intelligent Vehicles*, pages 311–342. Springer, London, 2012. DOI: 10.1007/978-0-85729-085-4. 83

L. Nunes and E. C. Oliveira. Learning from multiple sources. In N. R. Jennings, C. Sierra, L. Sonenberg, and M. Tambe, Editors, *Proceedings of the 3rd International Joint Conference on Autonomous Agents and Multi Agent Systems, AAMAS*, volume 3, pages 1106–1113, New York, USA, July 2004. New York, IEEE Computer Society. 69

D. de Oliveira and A. L. C. Bazzan. Multiagent learning on traffic lights control: effects of using shared information. In A. L. C. Bazzan and F. Klügl, Editors, *Multi-Agent Systems for Traffic and Transportation*, pages 307–321. IGI Global, Hershey, PA, 2009. ISBN 978-160566226-8. DOI: 10.4018/978-1-60566-226-8. 70

D. de Oliveira, A. L. C. Bazzan, and V. Lesser. Using cooperative mediation to coordinate traffic lights: a case study. In F. Dignum, V. Dignum, S. Koenig, S. Kraus, M. P. Singh, and M. Wooldridge, Editors, *Proceedings of the 4th International Joint Conference on Autonomous Agents and Multi Agent Systems (AAMAS)*, pages 463–470. New York, IEEE Computer Society, July 2005. 66

N. Oppenheim. *Urban Travel Demand Modeling: From Individual Choices to General Equilibrium*. John Wiley & Sons, Inc., New York, NY, January 1995. ISBN 0-471-55723-4. 25

J. de Dios Ortúzar and L. G. Willumsen. *Modelling Transport*. John Wiley & Sons, 3rd edition, 2001. 23, 24, 25, 27, 30, 42, 43

S. Panwei and H. Dia. A fuzzy neural approach to modeling behavioural rules in agent-based route choice simulations. In A. L. C. Bazzan, B. Chaib-Draa, F. Klügl, and S. Ossowski, Editors, *Proceedings of the 4th Workshop on Agents in Traffic and Transportation, AAMAS 2006, May, 9th, Hakodate, JP*, pages 70–78, 2006. 63

M. Papageorgiou. Traffic control. In R. W. Hall, editor, *Handbook of Transportation Science*, chapter 8, pages 243–277. Kluwer Academic Pub, 2003. 22

Z. Papp. Situation awareness in intelligent vehicles. In A. Eskandarian, editor, *Handbook of Intelligent Vehicles*, pages 61–80. Springer, London, 2012. DOI: 10.1007/978-0-85729-085-4. 78, 79

L. S. Passos, Z. Kokkinogenis, and R. J. F. Rossetti. Towards the next-generation traffic simulation tools: a first appraisal. *3rd Workshop on Intelligent Systems and Applications (WISA), 6th Iberian Conference on Information Systems and Technologies (CISTI'11)*, June 2011. 53

M. Patriksson. *The traffic assignment problem: models and methods*. VSP, Utrecht, The Netherlands, 1994. 43

S. Peeta and A. K. Ziliaskopoulos. Foundations of dynamic traffic assignment: The past, the present and the future. *Networks and Spatial Economics*, 1:233–265, 2001. DOI: 10.1023/A:1012827724856. 43

J. Piao and M. McDonald. Advanced driver assistance systems from autonomous to cooperative approach. *Transportation Reviews*, 28(5):659–6841–27, August 2008. DOI: 10.1080/01441640801987825. 80, 81

A. R. Pinjari, R. M. Pendyala, C. R. Bhat, and P. A. Waddell. Modeling the choice continuum: an integrated model of residential location, auto ownership, bicycle ownership, and commute tour mode choice decisions. *Transportation*, 38(6):933–958, 2011. DOI: 10.1007/s11116-011-9360-y. 30

L. A. Prashanth and S. Bhatnagar. Reinforcement learning with function approximation for traffic signal control. *IEEE Transaction on Intelligent Transportation Systems*, 12(2):412–421, June 2011. DOI: 10.1109/TITS.2010.2091408. 69

H. Prothmann, S. Tomforde, J. Branke, J. Hähner, C. Müller-Schloer, and H. Schmeck. Organic traffic control. In C. Müller-Schloer, H. Schmeck, and T. Ungerer, Editors, *Organic Computing – A Paradigm Shift for Complex Systems*, pages 431–446. Springer Basel, 2011. ISBN 978-3-0348-0130-0. http://dx.doi.org/10.1007/978-3-0348-0130-0_28. DOI: 10.1007/978-3-0348-0130-0. 68

M. A. Quddus, R. Noland, and W. Ochieng. A high accuracy fuzzy logic based map matching algorithm for road transport. *Journal of Intelligent Transportation Systems: Technology, Planning, and Operations*, 10(3):103–115, 2006. DOI: 10.1080/15472450600793560. 84

M. A. Quddus, W. Y. Ochieng, and R. B. Noland. Current map-matching algorithms for transport applications: State-of-the art and future research directions. *Transportation Research Part C: Emerging Technologies*, 15(5):312–328, 2007. DOI: 10.1016/j.trc.2007.05.002. 84

I. Radusch. Collaborative mobility - beyond communicating vehicles. *ERCIM News, Special issue on Intelligent Vehicles*, 94:4, July 2013. 95

B. Ran and D. E. Boyce. *Modeling dynamic transportation networks: an intelligent transportation system oriented approach.* Springer, 1996. ISBN 9783540583608. 38, 39

A. Rao and M. Georgeff. Modeling rational agents within a BDI architecture. In *Proceedings of the International Conference on Principles of Knowledge Representation and Reasoning*, pages 473–484, 1991. 64

F. Ricci. Travel recommender systems. *IEEE Intelligent Systems*, 17(6):55–57, 2002. 88

F. Ricci. Mobile recommender systems. *International Journal of Information Technology and Tourism*, 12:205–231, 2010. DOI: 10.3727/109830511X12978702284390. 88

G. Rindsfüser and F. Klügl. The scheduling agent - using sesam to implement a generator of activity programs. In H. Timmermans, editor, *Progress in Activity-Based Analysis*. Elsevier, 2005. 29

D. I. Robertson. TRANSYT: A traffic network study tool. Rep. LR 253, Road Res. Lab., London, 1969. 19, 20

M. Röckl, P. Robertson, K. Frank, and T. Strang. An architecture for situation-aware driver assistance systems. In *Proceedings of the 65th IEEE Vehicular Technology Conference*, pages 2555–2559. IEEE, 2007. DOI: 10.1109/VETECS.2007.526. 78, 79

R. P. Roess, E. S. Prassas, and W. R. McShane. *Traffic Engineering*. Prentice Hall, 3rd edition, 2004. 3, 10, 13, 16, 18, 21, 22, 45, 46

N. A. Ronald, T. Arentze, and H. Timmermans. An agent-based framework for modelling social influence on traveler behaviour. *18th World IMACS/MODSIM Congress*, pages 2955–2961, 2009. 64

D. A. Roozemond. Using intelligent agents for pro-active, real-time urban intersection control. *European Journal of Operational Research*, 131(2):293–301, June 2001. DOI: 10.1016/S0377-2217(00)00129-6. 66

G. Rose. Mobile phones as traffic probes: Practices, prospects and issues. *Transport Reviews*, 26 (3):275–291, 2006. DOI: 10.1080/01441640500361108. 96

R. J. F. Rossetti, R. H. Bordini, A. L. C. Bazzan, S. Bampi, R. Liu, and D. Van Vliet. Using BDI agents to improve driver modelling in a commuter scenario. *Transportation Research Part C: Emerging Technologies*, 10(5–6):47–72, 2002. DOI: 10.1016/S0968-090X(02)00027-X. 63

R. J. F. Rossetti, P. A. F. Ferreira, R. A. M. Braga, and E. C. Oliveira. Towards an artificial traffic control system. In *Proc. of the 11th International IEEE Conference on Intelligent Transportation Systems (ITSC 2008)*, pages 14–19, Beijing, China, October 2008. DOI: 10.1109/ITSC.2008.4732719. 75

T. Roughgarden and É. Tardos. How bad is selfish routing? *J. ACM*, 49(2):236–259, 2002. DOI: 10.1145/506147.506153. 42

J. Rybicki. *Cooperative Traffic Information Systems Based on Peer-to-Peer Networks*. Ph.D. thesis, Mathematisch-Naturwissenschaftlichen Fakultät der Heinrich-Heine-Universität Düsseldorf, 2011. 81

A. Salkham, R. Cunningham, A. Garg, and V. Cahill. A collaborative reinforcement learning approach to urban traffic control optimization. In *Proc. of International Conference on Web Intelligence and Intelligent Agent Technology*, pages 560–566. IEEE, 2008. DOI: 10.1109/WI-IAT.2008.88. 69

D. D. Salvucci, E. R. Boer, and A. Liu. Toward an integrated model of driver behavior in a cognitive architecture. *Transportation Research Record*, pages 9–16, 2001. DOI: 10.3141/1779-02. 63

H. Schepperle and K. Böhm. Agent-based traffic control using auctions. In M. Klusch, K. V. Hindriks, M. P. Papazoglou, and L. Sterling, Editors, *Proc. of the CIA*, pages 119–133. Springer, 2007. 73

H. Schepperle and K. Böhm. Valuation-aware traffic control – the notion and the issues. In A. L. C. Bazzan and F. Klügl, Editors, *Multi-Agent Systems for Traffic and Transportation*, pages 218–239. IGI Global, Hershey, PA, 2009. ISBN 978-1-60566-226-8. DOI: 10.4018/978-1-60566-226-8. 73

B. Scheuermann, C. Lochert, J. Rybicki, and M. Mauve. A fundamental scalability criterion for data aggregation in VANETs. In *MobiCom '09: Proceedings of the 15th Annual ACM International Conference on Mobile Computing and Networking*, pages 285–296, 2009. DOI: 10.1145/1614320.1614352. 81

N. Schüssler and K. W. Axhausen. Identifying trips and activities and their characteristics from GPS raw data without further information. In *8th International Conference on Survey Methods in Transportation, Annecy, May 2008*, 2008. 89

S. Seele, R. Herpers, and C. Bauckhage. Cognitive agents for microscopic traffic simulations in virtual environments. In M. Herrlich, R. Malaka, and M. Masuch, Editors, *Proc. of ICEC 2012, LNCS 7522*, pages 318–325, 2012. 62

Sukhjit Singh Sehra, Jaiteg Singh, and Hardeep Singh Rai. Assessment of OpenStreetMap data - a review. *International Journal of Computer Applications*, 76(16):17–20, August 2013. DOI: 10.5120/13331-0888. 51

J. Sewall, D. Wilkie, and M. C. Lin. Interactive hybrid simulation of large-scale traffic. *ACM Transaction on Graphics (Proceedings of SIGGRAPH Asia)*, 30(6), December 2011. DOI: 10.1145/2070781.2024169. 59

Y. Sheffi. *Urban Transportation Networks: Equilibrium Analysis With Mathematical Programming Methods*. Prentice-Hall, 1985. ISBN 9780139397295. 43

B. Castro da Silva, E. W. Basso, A. L. C. Bazzan, and P. M. Engel. Dealing with non-stationary environments using context detection. In W. W. Cohen and A. Moore, Editors, *Proceedings of the 23rd International Conference on Machine Learning ICML*, pages 217–224. New York, ACM Press, June 2006. DOI: 10.1145/1143844. 69

H. A. Simon. *Models of the Man*. J. Wiley & Sons, New York, 1957. 40

I. Skog and P. Handel. State-of-the-art in-car navigation: An overview. In A. Eskandarian, editor, *Handbook of Intelligent Vehicles*, pages 435–462. Springer, London, 2012. DOI: 10.1007/978-0-85729-085-4. 83

M. Steingröver, R. Schouten, S. Peelen, E. Nijhuis, and B. Bakker. Reinforcement learning of traffic light controllers adapting to traffic congestion. In K. Verbeeck, K. Tuyls, A. Nowé, B. Manderick, and B. Kuijpers, Editors, *Proceedings of the Seventeenth Belgium-Netherlands Conference on Artificial Intelligence (BNAIC 2005)*, pages 216–223, Brussels, Belgium, October 2005. Koninklijke Vlaamse Academie van Belie voor Wetenschappen en Kunsten. 71

A. Stevens, A Quimby, A Board, T Kerslott, and P. Burns. Design guidelines for safety of in-vehicle information systems, 2002. 79

P. R. Stopher, C. Prasad, L. Wargelin, and J. Minser. Conducting a GPS-only household travel survey. In J. Zmud, M. Lee-Gosselin, M. Munizaga, and J. Antonio Carrasco, Editors, *Transport Survey Methods: Best practice for Decision Making*, chapter 5, pages 91–114. Emerald Group Publishing, Bingley, UK, 2013. 89

J. Sussman. *Introduction to transportation systems*. Artech House, 2000. xiv, 22

A. R. Tavares and A. L. C. Bazzan. A multiagent based road pricing approach for urban traffic management. In *Third Brazilian Workshop on Social Simulation*, pages 99–105, 2012. DOI: 10.1109/BWSS.2012.22. 73

N. B. Taylor. The CONTRAM dynamic traffic assignment model. *Networks and Spatial Economics*, 3:297–322, 2003. DOI: 10.1023/A:1025394201651. 59

H. Timmermans, editor. *Progress in Activity-based Analysis*. Elsevier, 2005. 30

H. Timmermans, T. Arentze, and C.-H. Joh. Analysing space-time behaviour: new approaches to old problems. *Progress in Human Geography*, 26:175–190, 2002. DOI: 10.1191/0309132502ph363ra. 29

K. Train. *Discrete Choice Methods with Simulation*. Cambridge University Press, 2003. DOI: 10.1017/CBO9780511753930. 30

*Highway Capacity Manual*. Transportation Research Board, 2001. 10, 45

TRANSYT-7F. *TRANSYT-7F User's Manual*. Transportation Research Center, University of Florida, 1988. 20

M. Treiber and A. Kesting. *Traffic Flow Dynamics*. Springer Berlin Heidelberg, 2013. ISBN 9783642324604. DOI: 10.1007/978-3-642-32460-4. 22, 48, 51, 54, 56, 58

M. Treiber, A. Hennecke, and D. Helbing. Congested traffic states in empirical observations and microscopic simulations. *Phys. Rev. E*, 62:1805–1824, 2000. DOI: 10.1103/PhysRevE.62.1805. 56

T. Tsekeris and S. Voß. Design and evaluation of road pricing: state-of-the-art and methodological advances. *NETNOMICS: Economic Research and Electronic Networking*, 10(1):5–52, 2009. ISSN 1385-9587. DOI: 10.1007/s11066-008-9024-z. 22, 75

K. Tumer and D. Wolpert. Collective intelligence and Braess' paradox. In *Proceedings of the Seventeenth National Conference on Artificial Intelligence*, pages 104–109. AAAI Press, 2000. 41

K. Tumer, Z. T. Welch, and A. Agogino. Aligning social welfare and agent preferences to alleviate traffic congestion. In Lin Padgham, David Parkes, J. Müller, and Simon Parsons, Editors, *Proceedings of the 7th Int. Conference on Autonomous Agents and Multiagent Systems*, pages 655–662, Estoril, May 2008. IFAAMAS. 72

K. Tumer, Z. T. Welch, and A. Agogino. Traffic congestion management as a learning agent coordination problem. In A. L. C. Bazzan and F. Klügl, Editors, *Multi-Agent Systems for Traffic and Transportation*, pages 261–279. IGI Global, Hershey, PA, 2009. ISBN 978-1-60566-226-8. DOI: 10.4018/978-1-60566-226-8. 72

Y. Umemura. Driver behavior and active safety. *R & D Review of Toyota CRDL - Special Issue: Driver Behavior and Active Safety*, 39(2), 2004. 78, 79

M. van den Berg, A. Hegyi, B. de Schutter, and J. Hellendoorn. A macroscopic traffic flow model for integrated control of freeways and urban traffic networks. In *Proc. of the 42th IEEE Conference on Decision and Control, Maui, Hawaii, Dec. 2003*, pages 2774–2779, 2003. DOI: 10.1109/CDC.2003.1273044. 59

R. van Katwijk and P. van. Koningsbruggen. Coordination of traffic management instruments using agent technology. *Transportation Research Part C: Emerging Technologies*, 10(5-6):455 – 471, 2002. ISSN 0968-090X. DOI: 10.1016/S0968-090X(02)00034-7. 74

R.T. van Katwijk, P. van Koningsbruggen, B. De Schutter, and J. Hellendoorn. A test bed for multi-agent control systems in road traffic management. In F. Klügl, A. L. C. Bazzan, and S. Ossowski, Editors, *Applications of Agent Technology in Traffic and Transportation*,

Whitestein Series in Software Agent Technologies and Autonomic Computing, pages 113–131. Birkhäuser, Basel, 2005. DOI: 10.1007/3-7643-7363-6. 74

M. Vasirani and S. Ossowski. A market-based approach to reservation-based urban road traffic management. In K. Decker, J. Sichman, C. Sierra, and C. Castelfranchi, Editors, *Proc. of the 8th Int. J. Conf. on Autonomous Agents and Multiagent Systems (AAMAS)*, pages 617–624, Budapest, May 2009. IFAAMAS. 73

M. Vasirani and S. Ossowski. A computational market for distributed control of urban road traffic systems. *IEEE Trans. on Int. Transp. Systems*, 12(2):313–321, June 2011. ISSN 1524-9050. DOI: 10.1109/TITS.2011.2104956. 73, 91

M. Wachs. Policy implications of recent behavioral research in transportation demand management. *Journal of Planning Literature*, 5(4):333–341, 1991. DOI: 10.1177/088541229100500402. 32

J. Wahle, L. Neubert, J. Esser, and M. Schreckenberg. A cellular automaton traffic flow model for online simulation of traffic. *Parallel Computing*, 27(5):719–735, 2001. DOI: 10.1016/S0167-8191(00)00085-5. 57

Fei-Yue Wang. Toward a revolution in transportation operations: AI for complex systems. *IEEE Intelligent Systems*, 23(6):8–13, 2008. ISSN 1541-1672. DOI: 10.1109/MIS.2008.112. 75

N. Wanichayapong, W. Pruthipunyaskul, W. Pattara-Atikom, and P. Chaovalit. Social-based traffic information extraction and classification. In *11th International Conference on ITS Telecommunications (ITST)*, pages 107–112, 2011. DOI: 10.1109/ITST.2011.6060036. 96

J. G. Wardrop. Some theoretical aspects of road traffic research. *Proceedings of the Institute of Civil Engineers*, 1(3):325–362, 1952. 35

D. Watling. Stability of the stochastic equilibrium assignment problem: a dynamical systems approach. *Transportation Research Part B: Methodological*, 33(4):281–312, 1999. DOI: 10.1016/S0191-2615(98)00033-2. 42

M. Wiering. Multi-agent reinforcement learning for traffic light control. In *Proceedings of the Seventeenth International Conference on Machine Learning (ICML 2000)*, pages 1151–1158, 2000. 71

P. Winters. Transportation demand management (millennium papers; a5010: Committee on transportation demand management), November 1999. 32

M. J. Wooldridge. *An Introduction to MultiAgent Systems*. John Wiley & Sons, Chichester, 2009. 2nd edition. 64

C. C. Wright. Control of drivers' route choice: Pipe dream or panacea? *Transportation*, 7:193–210, 1978. DOI: 10.1007/BF00184639. 82

S. Wright, N. J. Ward, and Anthony G. Cohn. Enhanced presence in driving simulators using autonomous traffic with virtual personalities. *Presence*, 11(6):578–590, July 2002. DOI: 10.1162/105474602321050712. 62

T. Yamashita and K. Kurumatani. New approach to smooth traffic flow with route information sharing. In A. L. C. Bazzan and F. Klügl, Editors, *Multi-Agent Systems for Traffic and Transportation*, pages 291–306. IGI Global, 2009. DOI: 10.4018/978-1-60566-226-8. 86

Yang Yang, Xu Li, Wei Shu, and Min-You Wu. Quality evaluation of vehicle navigation with CPS. In *GLOBECOM '10: Proceedings of the IEEE Global Communications Conference*, 2010. DOI: 10.1109/GLOCOM.2010.5683433. 81

J.Y. Yen. Finding the k shortest loopless paths in a network. *management Science*, pages 712–716, 1971. DOI: 10.1287/mnsc.17.11.712. 35

Y. Yim. The state of cellular probes, 2003. http://www.path.berkeley.edu/PATH/Publications/PDF/PRR/2003/PRR-2003-25.pdf 96

C. Zhang, S. Abdallah, and V. Lesser. Integrating organizational control into multi-agent learning. In J. S. Sichman, K. S. Decker, C. Sierra, and C. Castelfranchi, Editors, *Proceedings of the 8th International Conference on Autonomous Agents and Multiagent Systems (AAMAS)*, pages 757–764, Budapest, Hungary, 2009. 70

Jiangping Zhou and Shuai Dai. Urban and metropolitan freight transportation: A quick review of existing models. *Journal of Transportation Systems Engineering and Information Technology*, 12(4):106–114, 2012. DOI: 10.1016/S1570-6672(11)60214-6. 32

D. Zielstra and H. H. Hochmair. Using free and proprietary data to compare shortest-path lengths for effective pedestrian routing in street networks. *Journal of the Transportation Research Board*, 2299(1):41–47, 2012. DOI: 10.3141/2299-05. 51

# Authors' Biographies

## ANA L. C. BAZZAN

**Ana Bazzan** received her Ph.D. in 1997 from the University of Karlsruhe, Germany. Currently, she is an associate professor at UFRGS (Computer Science Department) in Porto Alegre, Brazil. She is Associate Editor of the journals *Autonomous Agents and Multiagent Systems*, *Advances in Complex Systems*, and *Journal of Multiagent and Grid Systems*, and co-general chair of the AA-MAS 2014 conference. Her research interests include: multiagent systems, multiagent learning, complex systems, agent-based simulation, and applications of AI and multiagent techniques in traffic simulation and control.

## FRANZISKA KLÜGL

**Franziska Klügl** has been a full professor in Information Technology at Örebro University, Sweden since 2011. Before that she was senior lecturer at Örebro University and assistant professor at the University of Würzburg, Germany. She holds a Ph.D. (2000) and the Venia Legendi (2009) in Computer Science from the University of Würzburg. Her main areas of interests are methodologies, tools, and applications of multi-agent systems, mainly in simulation. Her main application area is agent-based traffic simulation.

Printed in the United States
by Baker & Taylor Publisher Services